PRINCIPLES AND PRACTICE

Springer

*Berlin
Heidelberg
New York
Barcelona
Hong Kong
London
Milan
Paris
Singapore
Tokyo*

Paul Popescu Hélène Hayes Bernard Dutrillaux (Eds.)

Techniques in Animal Cytogenetics

With 74 Figures, 7 in Colour

Springer

Prof. Dr. Paul Popescu
Directeur de Recherche à l'INRA
Dr. Helene Hayes
Chargée de Recherche
Laboratoire de Génétique Biochimique et de Cytogénétique
Institut National de la Recherche Agronomique (CRJ)
Domaine de Vilvert – Bat 440
78352 Jouy-en-Josas Cedex, France

Dr. Bernard Dutrillaux
Directeur de Recherche au CNRS
Institut Curie
CNRS UMR 147
26, rue d'Ulm
75231 Paris Cedex 05, France

Translated from the French edition by Ruxandra Popescu

Techniques de cytogénétique animale (Techniques et Pratiques), P. Popescu, H. Hayes, B. Dutrillaux, coord. INRA, Paris 1998 (ISBN 2-7380-0819-4)
Cover illustration: Immunofluorescent revelation of 5-mC rich bands on human chromosomes. Photograph provided by C. Bourgeois.

ISBN 3-540-66737-7 Springer-Verlag Berlin Heidelberg New York

Library of Congress Cataloging-in-Publication Data
Techniques de cytogénétique animale. English.
     Techniques in animal cytogenetics / Paul Popescu, Hélène Hayes, Bernard Dutrillaux (eds.).
     p. cm. – (Principles and practice)
     Includes bibliographical references (p. 209).
     ISBN 3540667377 (hard : alk. paper)

Springer-Verlag Berlin Heidelberg New York
a member of BertelsmannSpringer Science & Business Media GmbH

© Springer-Verlag Berlin Heidelberg 2000, INRA Paris
Printed in Germany

Cover design: design & production GmbH, Heidelberg
Typesetting: Best-set Typesetter Ltd., Hong Kong
Printed on acid free paper SPIN   10660543   31/3130 as   5 4 3 2 1 0

Preface

A better "casting" could not be conceived. The authors of this book are gold-smiths on the subject. I have followed their work since their "entry" into cytogenetics and I have a high esteem for them. I consider it an honour to be asked to write the preface of their opus.

Paul Popescu, Directeur de Recherche at INRA, has also played a prominent part in the development of animal cytogenetics, especially in domestic animals. He is able to tell you the cost of a translocation in a pig breeding farm or a cow population: a fortune! P. Popescu has played a great part in gene mapping of these species using "in situ DNA hybridisation". His contributions are recognised world-wide. His laboratory receives many visitors every year and it serves as a reference for domestic animal cytogenetics.

Hélène Hayes, Chargé de Recherche at INRA, has collaborated with P. POPESCU in the elaboration of the "at hand" techniques and in many other discoveries which are listed in her bibliography. She showed the fascinating correspondence between bovine and human chromosomes and the compared gene maps of domestic bovidae.

Bernard Dutrillaux, Directeur de Recherche at CNRS, pursued his first campaign in the laboratory of Professor Jérôme Lejeune with whom he played a great part in the development of human cytogenetics. Then, he "emigrated" to found his own laboratory at Institut Curie. His work is recognised world-wide: the perfection of many chromosome marking techniques; a study of meiotic chromosomes; the elaboration of primate chromosomal phylogeny and of many other phylums (canides, pinnipedes, rodents, etc.)...a notable work indeed; a major contribution to the knowledge of the role of chromosome rearrangements at the onset and during the evolution of oncogenesis; the development of the oncogene and antioncogene concept. The work of B. Dutrillaux and his associates represents an impregnable bastion of human and animal cytogenetics.

After the first research in Drosophila and mouse, cytogenetics expanded rapidly with Tjio and Levan (1954) who demonstrated that the human chromosome number is 46, not 48 as was believed before. Then, J. Lejeune discovered trisomy 21. A real explosion of discoveries in congenital human pathology and in cancerology succeeded thereafter. Cytogenetics allowed gene mapping of man and of many animal species; deciphering the phylogeny of primates and many species; the understanding of the genome function; deciphering the part played by the lampbrush chromosomes; the identification of embryonic development control; in vitro fertilisation; in utero diagnosis; transgenic techniques; gene therapy.

This extraordinary explosion of genetics was rendered possible thanks to the perfection of a load of specific techniques which are the subject of this handbook. It is, of course, a "livre de recettes". But thanks to them, it represents also as many open windows on the achievements they made possible: obtaining chromosome spreads, staining and high-resolution chromosome labeling; in situ hybridisation leading to genic localisations; meiotic and embryonic studies; interspecific in vitro fertilization and ascertainment of the prevalence of sperm aneuploidies; mapping of lampbrush chromosomes in pleurodeles, of Drosophila polytenic chromosomes; sex chromatin staining; flow-cytometry and chromosome sorting.

All these techniques adapted to the animal field have their counterpart in vegetal cytogenetics (see the book of Jahier "Techniques de Cytogénétique Végétale").

It stands to reason that the "Popescu-Hayes-Dutrillaux" will have a long and brilliant road by the same virtue as the ever existing "Ali-Bab" in the art of cooking (first edition 1928). It is a "must" for the laboratory and the library of biologists to whom cytogenetics is an essential ally: geneticists, zoologists, gene cartographers, oncologists, embryologists. And the list is far from finished. It will be a wonderful thesis-present for those who undertake to penetrate the secrets of life; they are still numerous.

Dr. Jean de GROUCHY
Directeur de Recherche Honoraire au CNRS

List of Contributors

S. Aulard
CNRS-Populations, Génétique et Evolution-UPR 9034
91198 Gif-sur-Yvette Cedex
France

C.A. Bourgeois
Institut Curie
CNRS UMR 147
26, rue d'Ulm
75231 Paris Cedex 05
France

D. Celeda
Institut für präventive Biomedizin
Hauptstrasse 82
67433 Neustadt
Germany

C. Cremer
Ruprecht Karls Universität Heidelberg
Institut für Angewandte Physik
Albert Ueberle Str. 3–5
69120 Heidelberg
Germany

B. Dutrillaux
Institut Curie
CNRS UMR 147
26, rue d'Ulm
75231 Paris Cedex 05
France

F.J. Ectors
Faculté de Médecine Vétérinaire
Département d'Embriologie et Reproduction
Université de Liège
B-4000 Liège
Belgique

O. Gabriel-Rodez
Institut d'Embriologie
Faculté de Médecine
11, rue Humann
67085 Strasbourg Cedex
France

M. Hausmann
Institut für Molekulare Biotechnologie e.V.
Wissenschaftsgemeinschaft Gottfried Wilhelm Leibniz e.V.
Postfach 100813
D-07708 Jena
Germany

H. Hayes
INRA-Unité de Génétique biochimique et cytogénétique
78352 Jouy-en-Josas Cedex
France

L. Koulischer
Centre de Génétique
CHU Sart-Tilman
4000 Liège
Belgique

J.C. Lacroix
Laboratoire de Génétique du Développement
Université Pierre et Marie Curie
9, Quai St. Bernard-Bat. C
75005 Paris
France

F. Lemeunier
CNRS-Populations, Génétique et Evolution-UPR 9034
91198 Gif-sur-Yvette Cedex
France

P. Popescu
INRA-Unité de Génétique biochimique et cytogénétique
78352 Jouy-en-Josas Cedex
France

C. Pyne
Laboratoire de Génétique du Développement
Université Pierre et Marie Curie
9, Quai St. Bernard-Bat. C
75005 Paris
France

Y. RUMPLER
Institut d'Embriologie
Faculté de Médecine
11, rue Humann
67085 Strasbourg Cedex
France

F. SIMON
Laboratoire de Génétique du Développement
Université Pierre et Marie Curie
9, Quai St. Bernard-Bat. C
75005 Paris
France

N. VICHNIAKOVA
Laboratoire de Génétique du Développement
Universtité Pierre et Marie Curie
9, Quai St. Bernard-Bat. C
75005 Paris
France

Abbreviations

5-ACR	5-azacytidine
5-dACR	5-deoxyazacytidine
5-mC	5-methylcytosine
A	adenine
AT	adenine-thymine base pair
BrdU	5-bromo-2′-deoxyuridine
BSA	bovine serum albumine
BSS	balanced salt solution
C	cytosine
CCD	cooled camera device
CI	centromeric index
COC	cumulus – ovocyte complexe
ConA	concanavalin A
DA	distamycin A
DAB	3,3′-diaminobenzidine tetrahydrochloride
DAPI	4′-6-diamino-2-phenylindole
dATP	2′-deoxyadenosine 5′-triphosphate
dCTP	2′-deoxycytidine 5′-triphosphate
DMSO	dimethylsulfoxide
DNA	deoxyribonucleic acid
DNase	deoxyribonuclease
dNTP	2′-deoxynucleotide 5′-triphosphate
DTT	dithiothreitol
dTTP	2′-deoxythymidine 5′-triphosphate
dUDP	deoxyuridine triphosphate
EDTA	ethylene diamine tetra-acetic acid
ETB	ethidium bromide
FCS	foetal calf serum
FdU	5-fluoro-2′-deoxyuridine
FISH	fluorescent in situ hybridization
FITC	fluorescein isothiocyanate
FPG	fluorochrome-photolysis-Giemsa
FSH	follicle stimulating hormone

G	guanine
GC	guanine-cytosine base pair
I.U	international units
INRA	Institut National de Recherche Agronomique
ISCN	"International System for human Cytogenetics Nomenclature"
ISCNDA	International System for Cytogenetic Nomenclature of Domestic Animals
kb	kilobase
LH	luteinizing hormone
LINEs	long interspersed repeated sequences
LPS	lipopolysaccharide
MTX	methotrexate
NOR	nucleolar organising region
bp	base pair
PBS	phosphate buffered saline
PCR	polymerase chain reaction
PHA	phytohemagglutinin
PI	propidium iodide
PMT	photomultiplier
PPD	p-phenylenediamine
PVP	polyvinylpyrrolidone
PWM	pokeweed mitogen
RNA	ribonucleic acid
Rnase	ribonuclease
SAR	scaffold associated regions
SCE	sister chromatid exchange
SDS	sodium dodecyl sulfate
SINEs	short interspersed repeated sequences
cpm	count per minute
SSC	sodium saline citrate
T	thymine
U.V.	ultraviolet light
vol.	volume
vs	versus
YAC	yeast artificial chromosome

NB. The Appendix contains the protocols for the solutions used throughout the book and is organised in 9 sub-chapters according to the type of solution.

S	Solutions for Chromosome Staining and Banding Techniques
M	Miscellaneous
L	Solutions for Lampbrush Chromosomes
F	Solutions for Flow and Slit-Scan Cytometry
H	Solutions for In Situ Hybridisation
M	Culture Media
SS	Stock Solutions

B   Buffer Solutions
V   Solutions for In Vivo Treatements

Throughout the book, appendix, followed by a capital and a number, i.e. appendix S1, refers to the corresponding sub-chapter of the Appendix.

Nomenclature

P. POPESCU

The first conference for the standardisation of human chromosomes was held in Denver in 1960. Since then, several conferences of experts have improved the nomenclature and the classification of human chromosomes. The last conference was held in 1995 (ICSN, 1995).

This nomenclature defines the normal and pathologic formula and introduces the symbols for the most typical different numeric and structural abnormalities. These formula also contain a bands nomenclature for each chromosome. These bands are revealed by different banding methods. Karyotypes images and their corresponding ideograms are proposed as models. A detailed nomenclature knowledge is presented in the proceedings of ISCN (1978) and ISCN (1981). High resolution bands are published in ISCN (1995), which defines the symbols for fluorescent in situ hybridisation on chromosomes, decondensed chromatin and the interphasic nucleus.

The first international conference for the standardisation of domestic animals took place in Reading in 1976, published in 1980 (Reading Conference, 1976). It established G band standard karyotypes for the main domestic species: cattle, sheep, goat, horse and rabbit.

During the second conference for the standardisation of domestic animals, organised in Jouy-en-Josas/France in 1989, the cattle karyotype was described in Q, G and R bands and it was possible to construct ideograms using sequential Q and R banding. Goat and sheep karyotypes were described in R bands (ISCNDA, 1989). The standard karyotype of the horse was also published separately in 1989 (Richer et al., 1990).

From the conference held in Jouy-en-Josas, the standard karyotype of the sheep was improved during the 8th North American colloquium on domestic animals cytogenetics, held at Guelph University in 1993 (Ansari et al., 1994) and the 9th North American colloquium on cattle, held at Texas University, College Station in 1995 (Popescu et al., 1996).

The G bands pig karyotype, standardised during the Reading conference, was improved during the European colloquiums of domestic animals cytogenetics held in Uppsala (1980), Milan (1982) and Zürich (1984). In 1988 the standardisation Committee published a G band and R band standard karyotype of the pig and the corresponding ideograms (Committee of the Standardisation of the Domestic Pig, 1988).

In other species of domestic animals, only the karyotype of water buffalo (Ianuzzi, 1994) and dog (Switonski et al., 1996) have been standardised.

Mouse karyotype, was standardised in 1972 (Committee of Standardised Genetic Nomenclature for Mice, 1972) and that of the rat, in 1974 (Committee for a Standard Karyotype of *Rattus norvegicus,* 1973), as laboratory animals.

In spite of an important number of wild mammals species, probably about 2000 whose karyotype is known, only the karyotype of four species was standardised: 2 species of fox, the silver fox (*Vulpes fulvus)* (Mäkinen, 1985a) and the blue fox *(Alopex lagopus)* (Mäkinen, 1985b), the common shrew-mouse (Searle et al., 1991), and a rodent, the Peromyscus of North America (Greenbaum et al., 1994).

Contents

I	**PREPARATION OF CHROMOSOME SPREADS**	1
I.1	**Cell culture techniques**	
	H. HAYES, B. DUTRILLAUX	1
I.1.1	**Lymphocyte culture**	1
	Principle	1
	Protocol for whole blood	1
	Protocol for lymphocytes isolated by sedimentation	2
	Protocol for lymphocyte isolation by density gradient centrifugation	2
	Remarks	2
I.1.2	**Fibroblast cultures**	3
I.1.2.1	Cultures established from tissue fragments	3
	Principle	3
	Protocol	3
	Remarks	5
I.1.2.2	Initiation of cell cultures from bird embryos	5
	Principle	5
	Protocol	5
	Remarks	5
I.1.2.3	Cell counting	5
I.1.2.4	Cell storage	6
	Principle	6
	Protocol	7
I.2	**Preparation of chromosomes in prophase or prometaphase**	
	H. HAYES, B. DUTRILLAUX	7
I.2.1	**Single or double synchronisation using thymidine**	8
	Principle	8
	Protocol for lymphocytes (single synchronisation)	8
	Protocol for fibroblasts (double synchronisation)	8
	Remarks	10
I.2.2	**Synchronisation by amethopterin**	10
	Principle	10
	Protocol	10
	Remarks	11
I.2.3	**Synchronisation by 5-fluorodeoxiuridine**	

	P. Popescu, B. Dutrillaux	11
	Principle	11
I.2.4	Synchronisation by 5-bromodeoxyuridine or BrdU	11
	Principle	11
I.3	**Direct techniques and very short term development cultures**	12
I.3.1	Study of the bone marrow	12
	Protocol	12
	Remarks	12
I.3.2	Study of bird chromosomes from feather pulp	12
	Protocol	13
	Remarks	13
I.3.3	Study of fish chromosomes	13
	Protocol for culture of blood lymphocytes	14
	Protocol for culture of lymphocytes from kidney or spleen	14
	Protocol for chromosomal banding by BrdU incorporation in live fish	15
	Remarks	15
I.3.4	Study of insect chromosomes	16
I.3.4.1	Preparation of metaphase spreads from embryo cells	16
	Protocol	16
I.3.4.2	Preparation of metaphase spreads from gonads	16
	Protocol	17
I.3.4.3	Chromosome analysis from other tissus	17
I.3.4.4	General comments	17
I.4	**Preparation of chromosome spreads**	
	H. Hayes	18
	Principle	18
	Protocol	18
	Protocol for wet slides	19
	Protocol for dry slides	19
	Remarks	19
I.5	**Staining techniques of chromosome spreads**	
	H. Hayes, B. Dutrillaux	20
I.5.1	Classical staining using Giemsa	20
	Principle	20
	Protocol	20
	Remarks	20
I.5.2	Orcein staining	20
	Protocol 1	21
	Protocol 2	21
	Remarks	21
I.5.3	Acridine orange staining	22
	Protocol	22
	Remarks	22
I.5.4	Propidium iodide or DAPI staining	24

II	**CHROMOSOME BANDING TECHNIQUES**	**25**
II.1	**Introduction**	
	H. HAYES	25
II.1.1	**Chromosome organisation**	25
II.1.2	**Euchromatin**	27
	Distribution of SINE and LINE sequences along chromosomes	28
	Differential arrangement of AT alignment in Q/G and R bands	28
II.1.2.1	"Structural" bands	29
	Q bands	29
	G bands	30
	R bands	30
II.1.2.2	"Dynamic" bands	31
II.1.3	**Code of chromosome banding techniques**	33
II.2	**Techniques based on DNA structure**	
	B. DUTRILLAUX, H. HAYES	33
II.2.1	**QFQ bands using quinacrine mustard**	34
	Principle	34
	Protocol	34
	Remarks	34
II.2.2	**GTG banding by trypsin**	34
	Principle	34
	Protocol	34
	Remarks	34
II.2.3	**GAG banding by "denaturation"**	36
	Principle	36
	Protocol Acid-Saline-Giemsa	37
	Protocol Alkaline-Saline-Giemsa	37
	Remarks	37
II.2.4	**RHG banding by thermal denaturation**	37
	Principle	37
	Protocol	38
	Remarks	38
II.2.5	**T bands (terminal)**	39
	Protocol	39
	Remarks	41
II.2.6	**Bands rich in 5-methylcytocine**	
	C. Bourgeois	41
II.2.6.1	Introduction	41
II.2.6.2	Immunofluorescent revelation of 5-mC rich bands	43
	Principle	43
	Protocol	44
	Protocol of denaturation using hydrochloric acid	44
	Protocol of denaturation using ultraviolet lamp radiation	44
	Remarks	45

II.3   **Banding techniques based on DNA replication**

B. DUTILLAUX, H. HAYES 46

II.3.1   R or G bands by incorporation of BrdU 46

Principle 46

II.3.1.1   Immunoflorescent detection of BrdU incorporation
in chromosomes 48

Principle 48

Protocol 48

Remarks 49

II.3.1.2   FPG staining technique (fluorochrome-photolysis-
Giemsa) 49

Principle 49

Protocol 51

Remarks 51

II.3.1.3   Propidium iodide staining technique 51

Principle 51

Protocol 51

Remarks 52

II.3.1.4   DAPI staining technique 53

Principle 53

Protocol 53

II.3.1.5   Preparation of chromosomes labeled with BrdU during
the second half of the S phase to produce R bands 54

Protocol 54

Remarks 54

II.3.1.6   Preparation of chromosomes labeled with BrdU during
the first half of the S phase to produce G bands 55

Protocol 55

Remarks 55

II.3.2   **Sister chromatid exchanges (SCE)** 55

Principle 55

Protocol 55

Remarks 56

II.3.3   **Asymmetrical incorporation of BrdU** 56

II.4   **Techniques of chromosome differentiation
based on DNA base composition**

B. DUTRILLAUX 57

II.4.1   **Treatment by 5-azacytidine or 5-azadeoxycytidine** 57

Principle 57

Protocol 57

Remarks 57

II.5   **Heterochromatin staining** 58

II.5.1   **CBG bands** 58

Principle 58

Protocol 58

Remarks 58

II.5.2	CT bands	61
	Principle	61
	Protocol	61
II.5.3	G11 bands	61
	Principle	61
	Protocol	61
	Remarks	62
II.5.4	DA-DAPI staining	62
	Principle	62
	Protocol	64
	Remarks	64
II.6	**Staining of nucleolar organiser regions NOR**	
	H. HAYES, B. DUTRILLAUX	65
	Protocol	65
	Remarks	65
II.7	**Techniques of sequential banding**	
	B. DUTRILLAUX	65
	Principle	65
	Protocol of sequential Q, R, and C banding	68
	Protocol of sequential Q and NOR banding, or R and NOR banding	68
III	*IN SITU* **HYBRIDISATION TECHNIQUES**	
	H. HAYES, B. DUTRILLAUX	69
III.1	**Introduction**	69
III.1.1	*In situ* hybridisation using radioactive probes	69
III.1.2	*In situ* hybridisation using non radioactive probes	70
III.1.3	Identification of hybridised chromosomes	71
III.2	**Methods**	72
III.2.1	**Preparation of chromosome spreads**	72
III.2.2	**DNA probe labelling**	73
III.2.2.1	Non radioactive labelling of long DNA probes (>1 kb)	73
	Principle	73
	Protocol for nick translation labelling	73
III.2.2.2	Non radioactive labelling of short DNA probes (00.25-1.5 kb)	74
	Principle	74
	Protocol	74
	Remarks	74
III.2.2.3	Radioactive labelling	75
	Protocol for the use of the nick translation method	75
III.2.3	**Pretreatment of the chromosome preparations using ribonuclease A**	75
	Principle	75
	Protocol	75

III.2.4	Chromosomal DNA denaturation	75
	Principle	75
	Protocol	76
III.2.5	Probe preparation, labelling and denaturation	76
III.2.6	*In situ* hybridisation	76
III.2.7	Posthybridisation washes	76
III.2.7.1	Radioactive probes	76
III.2.7.2	Non radioactive probes	76
III.2.8	Hybridisation signal detection and banding	77
III.2.8.1	Radioactive probes: autoradiographic detection	77
	Principle	77
	Protocol	78
III.2.8.2	Non radioactives probes: immunoreaction detection	79
	Protocol	79
III.2.9	Microscopy and photography	79
III.2.9.1	Radioactive probes	79
III.2.9.2	Non radioactives probes	79
III.3	**Remarks on other applications of *in situ* hybridisation**	80
IV	**METHODS OF GERM CELLS STUDY**	85
IV.1	**Meiosis in male**	
	P. Popescu	85
IV.1.1	Classical method	85
	Protocol	85
	Remarks	86
IV.1.2	Synaptonemal complex method	
	Y. Rumpler, O. Gabriel-Robez	86
	Principle	88
	Protocol	89
	Remarks	92
IV.2	**Meiosis in the mammalian female**	
	J.F. Ectors, L. Koulisher	94
	Principle	94
	Protocol	95
IV.3	**Study of the spermatozoa by interspecific in vitro fertilisation (insemination)**	
	P. Popescu	97
	Principle	97
	Protocol	97
	Protocol of capacitation in the cold state	99
	Protocol of capacitation in the warm state	99
	Remarks	102

V	THE LAMPBRUSH CHROMOSOMES OF AMPHIBIANS	
	F. Simon, N. Vichniakova, C. payne, J.C. Lacroix	103
V.1	**Introduction: The basis of lampbrush chromosomes mapping**	**103**
V.1.1	Organisation of the lampbrush chromosomes	103
V.1.2	Morphologic mapping of the lampbrush chromosomes	107
V.1.2.1	Morphological variations of genetic origin	107
V.1.2.2	Morphological variations of physiological origin	109
V.1.3	**Immunomorphological mapping**	109
V.1.4	**Maps of the lampbrush chromosomes of _Pleurodeles_**	112
V.1.4.1	The conditions of map development	112
V.1.4.2	The intraspecific maps	112
V.2	**Technique for the preparation of lampbrush chromosomes for light microscopy**	**114**
V.2.1	**Preparation of oocytes**	114
	Protocol	114
V.2.2	**Chromosome preparation for morphological studies**	115
	Protocol	115
V.2.3	**Chromosome immunolabelling**	118
	Protocol	118
V.3	**Preparation of lampbrush chromosomes for electron microscopy**	**120**
V.3.1	**Ultrastructural studies**	120
	Protocol	121
V.3.2	**High resolution immunolabelling**	121
	Principle	121
V.3.2.1	Pre-embedding immunolabelling	123
	Protocol	123
V.3.2.2	Post-embedding immunolabelling	124
	Protocol	124
V.4	**Preparation of mitotic chromosomes**	**124**
	Protocol	125
V.5	**Analysis of lampbrush chromosomes**	**125**
V.5.1	**Morphological markers**	128
V.5.1.1	Axial structures	128
V.5.1.2	Loops	129
V.5.2	**Immunolabelling**	130
V.5.3	**Mapping parameters**	131
V.5.3.1	Chromosome classification	131
V.5.2.2	Chromosome orientation	131
V.5.3.3	Marker localisation	131

V.5.3.4	Correspondences between the lampbrush chromosome maps	131
V.6	**The importance of the lampbrush chromosomes**	**133**
V.6.1	Detection and analysis of chromosomal rearrangements	133
V.6.2	Study of populations and evolution of chromosomes	133
V.6.3	Study of the transcriptional physiology *in situ*	136
VI	**TECHNIQUES FOR THE SIUDY OF DROSOPHILA CHROMOSOMES**	
	F. LEMEUNIER, S. AULARD	137
VI.1	**Mitotic chromosomes**	**137**
VI.1.1	Introduction	137
VI.1.2	Chromosome spreads preparation	138
	Principle	138
	Protocol	138
VI.1.3	Staining and banding of mitotic chromosomes: remarks	139
VI.1.4	*In situ* hybridisation	140
VI.2	**Polythene chromosomes**	**140**
VI.2.1	Introduction	140
VI.2.2	Structure	141
VI.2.3	Reference maps	142
VI.2.4	Chromosome preparation: Classical and molecular cytogenetic technique	145
VI.2.4.1	Breeding conditions of Drosophila	145
VI.2.4.2	Chromosome preparation: classical analysis	145
	Protocol	145
VI.2.4.3	Chromosome preparation: in situ hybridisation	146
	Principle	146
	Protocol	146
VI.2.5	*In situ* hybridisation	147
	Protocol	147
VI.2.6	Chromosome observation	149
VII	**TECHNIQUES FOR THE STUEY OF INTERPHASE NUCLEUS**	**151**
VII.1	**Sex chromatin examination**	
	P. POPESCU	151
	Principle	151
	Protocol	151
	Remarks	151
VII.2	**Released chromatin (Chromatin halo)**	**152**
	Protocol	152

VII.3   *In situ* hybridisation of interphasic nuclei

C. BOURGEOIS   153
    Principle   153
    Protocol   153
    Protocol for paraffin sections   153
    Protocol for nucleus and spread cells or slides
       with cell layers   154
    Remarks   155

VIII   **APPLICATION OF FLOW CYTOMETRY
AND SLIT-SCAN FLUOROMETRY IN
ANALYSIS AND SORTING OF MAMMALIAN
CHROMOSOMES**

M. HAUSSMANN, C. CREMER   157
VIII.1   **Introduction**   157
VIII.2   **Principles of the flow cytometry**   158
VIII.2.1 **Standard flow cytometry**   158
VIII.2.2 **Slit scan fluorometry**   159
VIII.2.3 **Flow sorting**   161
VIII.2.4 **Computing**   161
VIII.3   **Methods of chromosome preparation and
staining**   161
VIII.3.1 **Modified hexanediole method**   162
    References   162
    Protocol   162
    Remarks   163
VIII.3.2 **TAcCaM – method**   163
    References   163
    Protocol   163
    Remarks   163
VIII.3.3 **Methanol – acetic acid method**   164
    References   164
    Protocol   164
    Remarks   164
VIII.3.4 **Tris/MgCl$_2$/Triton X10000 method**   164
    References   164
    Protocol   164
VIII.3.5 **Polyamine method**   164
    References   164
    Protocol   164
VIII.3.6 **Modified polyamine method**   165
    References   165
    Protocol   165
VIII.3.7 **HEPES/MgSO$_4$ method**   165
    References   165
    Protocol   165

VIII.3.8   Modified HEPES method 166
         References 166
         Protocol 166
         Remarks 166
VIII.3.9   Fluorescein labelling (FITC) by *in situ* hybridisation
         suspension 166
         References 166
         Protocol 166
         Remarks 167
VIII.3.10   **Dyes and lasers** 167
VIII.4   **Measurements and evaluation in flow
cytometry** 167
VIII.4.1   **Flow karyotypes** 167
VIII.4.2   **Data evaluation of flow karyotypes** 171
VIII.4.3   **Slit scan measurements** 172
VIII.4.4   **Perspectives** 175
VIII.4.5   **Identification of the sorted chromosomes by GTG
banding**
         P. POPESCU 176
         Principle 176
         Protocol 177
         Remarks 177
VIII.5   **In situ hybridisation to chromosomes
in suspension**
         D. CELEDA 178
         Principle 178
         Protocol 180
         Remarks 186

Appendix 187
S   Solutions for chromosome staining and banding
     techniques 187
M   Miscellaneous 189
L   Solutions for lampbrush chromosomes 192
F   Solutions for flow and slit scan cytometry 193
H   Solutions for in situ hybridisation 196
CM   Culture media 197
SS   Stock solutions 199
B   Buffer solutions 201
V   Solutions for in vivo treatments 203

Glossary 205
References 209
Index 225

I
Preparation of Chromosome Spreads

H. HAYES and B. DUTRILLAUX

I.1
Cell culture techniques

Chromosome analysis requires the preparation of slides rich in suitable chromosome spreads. The technique is divided into three steps:

- culture of lymphocytes stimulated by a mitogen or primary culture of fibroblasts
- accumulation of cells in metaphase or prometaphase, the stages of the cell cycle during which the chromosomes are most easily distinguishable
- harvesting of the cells and preparation of chromosome spreads.

I.1.1
Lymphocyte culture

Principle
Blood, which is easy to collect, is the source of cells most used in human cytogenetics. In Mammals, B and T lymphocytes are the only blood cells which can be transformed into actively proliferating cells. This lymphoblastic transformation induced *in vivo* by antigens can be stimulated *in vitro* by different mitogens, for example lectins. There are several types of lectins (Sharon and Lis, 1989), among which those most frequently used to transform lymphocytes are phytohaemagglutinin (PHA), concavalin A (Con A) and pokeweed mitogen (PWM). Their mitogenic effects are different: PHA and Con A preferentially stimulate the T lymphocytes, while PWM stimulates T and especially B lymphocytes. Moreover, these lectins do not stimulate the same sub-populations of T lymphocytes. Lymphocytes can be cultured either from whole blood (Moorhead et al., 1960), or after isolation, by sedimentation or by density gradient centrifugation (Boyum, 1968).

Protocol for whole blood
Blood is collected from an easily accessible vein (jugular in cattle) in a sterile heparinized tube (vacutainer). For each blood sample, different culture media are used, in order to increase the probability of obtaining good quality slides. The collected blood (0.5 ml) is added to 8 ml of Ham F12, TC199 or

RPMI1640 medium (appendix CM1) supplemented with 10% to 20% foetal calf serum (FCS) and a mitogen ConA (Serva) or PWM (IBF) at a final concentration of 10 µg/ml. The cultures are incubated in closed tubes at 37 °C for 3 days in the presence of ConA and for 4 or 5 days in that of PMW.

Protocol for lymphocytes isolated by sedimentation

Blood is collected as above and the tubes containing the blood are placed at room temperature for 30 to 80 minutes. The red cells which are denser than the lymphocytes sediment towards the lower half of the tube leaving a lymphocyte rich plasma layer at the top which is collected and used for the culture. This cell suspension (0.5 ml) is added to 8 ml of the culture media described above and the cultures are incubated under the same conditions as for the procedure with whole blood. Nucleated cells are counted after trypan blue staining (appendix M3) in order to inoculate cells at a final density of about 0.5 to 1.10^6 cells/ml.

Protocol for lymphocyte isolation by density gradient centrifugation

Lymphocytes can also be separated by centrifuging blood through a mixture of Ficoll (an agent which agglutinates the red cells) and sodium metrizoate or sodium diatrizoate (high density agents), the special density and osmotic pressure characteristics of which permit higher yields of lymphocytes (Boyum, 1968). Such mixtures are commercialised by Pharmacia Chemicals (Ficoll-Paque) and by Nycomed (Lymphoprep).

Blood is collected in a tube containing sodium heparin (final concentration 5 to 10 U.I./ml) and diluted with two volumes of sterile PBS⁻ (appendix B1). 15–20 ml of diluted blood is carefully layered onto 7.5 ml of the separation mixture (Ficoll-Paque or Lymphoprep) in a sterile centrifugation tube. After centrifugation at 1500 g for 15 minutes at 20 °C, the red cells agglutinated by Ficoll have sedimented towards the bottom of the tube, whereas the lymphocytes are situated at the interface of the plasma layer and the separation mixture. The supernatant is removed in order to eliminate as much as possible of the plasma and the platelets, and the lymphocytes are collected, care being taken to avoid contamination with red cells, and washed three times with serum free medium (3 vol.) by centrifugation at 400 g for 5 minutes. They are then counted after trypan blue staining (appendix M3), and resuspended in complete mitogen containing culture medium at a concentration of 0.5–1.10^6 cells/ml and incubated as described above for whole blood.

Remarks

In general, better results (preparations rich in analysable chromosome spreads) are obtained with cultured human lymphocytes than with cultured lymphocytes of other mammalian species. For many mammalian species, results are not reproducible from one experiment to another and the mitotic index is often low. This lack of reproducibility is probably due to the fact that the number of lymphocytes present in the collected blood, and their susceptibility to stimulation by mitogens vary greatly depending on the state of health of the animal used. The mitotic index depends on the number of lymphocytes which can be specifically transformed by the mitotic agent employed. The composition of the lymphocyte pool also differs according to species and it is quite possible that the lectins used to stimulate human lymphocytes are not adapted to those of other animal species. According to Renshaw et al. (1977), PHA yields good results with human lymphocytes, but does not stimulate bovine lymphocytes; it also fails to stimulate ovine lymphocytes (Hayes, personal communication). At present, Con A and PWM are the mitogens most used for the stimulation of non human mammalian lymphocytes.

I.1.2
Fibroblast culture

I.1.2.1
Cultures established from tissue fragments

If lymphocyte cultures do not yield good results for the preparation of chromosome spreads, they can be replaced by primary cultures of fibroblasts, which give reproducible results. In primary cultures, fibroblasts (irregularly shaped or fusi-form) generally become predominant in comparison to cells with epithelial morphology (polyhedral shape) which start dividing earlier but cannot be maintained as long in culture.

Principle
Primary cultures can be established from tissue taken from adult animals or foetuses. Cell cultures derived from foetal tissue are more useful than those derived from adult tissues for two reasons:

- they have a higher mitotic index
- they have a larger potential number of generations, about 50 as against 20 for those for cells from adult tissues.

The techniques of preparation and maintenance of cell cultures are the same in both cases. All fibroblast rich tissues can be used and the tissue selected varies from culture to culture. Removal of a skin sample after shaving and thorough disinfection with alcohol is often the only method practicable on live animals. Tissues can also be collected post-mortem. The permissible interval between death of the animal and tissue collection varies according to animal size, small-sized animals becoming contaminated very quickly. In this case, after external disinfection, a deep and poorly irrigated tissue such as a muscle tendon is chosen. With this procedure a delay even of several days after death may not prevent the establishment of cell cultures.

Protocol
■ **Preparation of skin biopsies.** The skin is washed with antiseptic soap, shaved and swabbed with 70% alcohol. A small fragment of skin is removed

with a sterile instrument, either with a punch or a scalpel. A fragment a few mm long and 1 mm thick is sufficient to initiate a culture. It is transferred to a sterile tube containing the culture medium (appendix CM2) or BSS$^+$ buffer (appendix B2).

■ **Preparation of foetus.** Embryos of domestic animals such as pig, cattle, goat or sheep are obtained either directly from the slaughter-house, at different stages of gestation, or from a pregnant female slaughtered at one third of gestation. The foetuses are removed aseptically from the uterus in the laboratory, rinsed with sterile PBS$^-$ (appendix B1) and then transferred to and handled in a sterile laminar flow hood. The same technique can be used with other domestic species such as rabbit, cat, dog, mink, mouse etc. Fragments of about 1 cm^3 of several tissues (muscle, lung, kidney etc.) are removed and placed in sterile Petri dishes containing culture medium MEM-10% FCS (appendix CM2).

■ **Tissue culture.** Each skin or foetal tissue fragment is placed in a small sterile beaker and cut with sterile scissors into pieces as small as possible which are then transferred to a flask containing a solution of 0.05% trypsin and 0.02% EDTA (appendix M13). The suspension is stirred, avoiding frothing, for 10 to 20 minutes at 37 °C to dissociate the cells. After decantation, the suspension of dissociated cells is collected and transferred to a 50 ml tube and 2.5 ml of foetal calf serum is added to inactivate trypsin. One or two further cycles of trypsin treatment can be carried out to recover as many cells as possible. The different suspensions are then combined, filtered through two or three layers of sterile gauze and the filtrate is centrifuged at 400 g for 5 minutes. The cell pellet is resuspended in 10 ml of culture medium and after staining with acetic violet (appendix M2) which lyses the red cells, the number of cells is counted using a Malassez chamber. This count is approximate because much organic debris remains mixed with the cells. The cells are then, either inoculated into medium MEM-10% FCS (appendix CM2) at a concentration of 0.5–3.10^6/ml, or frozen in a mixture of foetal calf serum and 10% DMSO (appendix V1) at a concentration of 10^7–10^8/ml and stored in liquid nitrogen for later use. The cultured cells are incubated at 37 °C in a 5% CO$_2$ atmosphere (these conditions are used in all further steps) for 24 hours to allow viable cells to adhere to the bottom of the flask; these are then washed once or twice with the medium to eliminate organic debris and dead cells. The culture medium is then renewed twice a week and when the cells are actively dividing and form foci of about one hundred cells, a trypsin treatment is performed to dissociate the latter. For this purpose, the culture medium is removed, the cells are rinsed with PBS$^-$ (appendix B1), 2 ml of a solution of 0.05% trypsin and 0.02% EDTA (appendix M13) is added and the cells are incubated for about 10 minutes at 37 °C. When the cells are completely detached and disaggregated, the trypsin is inactivated by the addition of 5 ml of medium MEM-10% FCS. The cell suspension is collected and centrifuged for 5 minutes at 400 g and the cell pellet is resuspended in 5 ml of culture medium, counted using a Malassez chamber (as described in paragraph I.1.2.3) without treatment and, then distributed into new flasks (75 cm^2) at a concentration of 1–2.10^5/ml.

■ **Cell subculture.** When the cells become confluent, it is necessary to collect them via a trypsin treatment and to subculture them as described above. This operation is referred to as one passage.

Remarks
Different foetal tissue cells can be cultured. However for pig, domestic cattle, sheep and goat, it has been frequently observed that lung cells give the best results. These cells can be maintained in culture for approximately 15 to 20 passages, but after 7 to 10 passages, they become less active and their mitotic index decreases. In cultures of adult skin cells, the total number of passages does not exceed 6 to 8 and only the first 3–4 passages are suitable for the preparation of chromosome spreads. This type of culture is therefore used only in special situations. The quality of the foetal calf serum added to culture media is very important for successful cell culture.

I.1.2.2
Initiation of cell cultures from bird embryos

Principle
The results of cultures of whole blood or lymphocytes from birds are less satisfactory, than those obtained with human or mammalian material, because of a low mitotic index. Primary cultures of fibroblast cells or, in particular of embryonic cells give better results.

Protocol
After incubation at 37.5 °C for 8–10 days, fertile eggs are broken in a sterile Petri dish, under a laminar flow hood. Organs such as the lung, heart or kidney are collected whole and also muscle as fragments. The rest of the procedure is carried out according to the protocol described in paragraph I.1.2.1.

Remarks
Bird embryo cells, like those from mammalian embryos, can be maintained in culture for approximately 15 passages, but the best chromosome preparations (Figure I.1) are obtained from the cells maintained for 2 to 7 passages.

I.1.2.3
Cell counting

Cells are counted using a Malassez chamber which is divided by a grid into 25 square compartments each containing 10^{-5} cm^3. A drop of the cell suspension, stained or not is placed between the chamber and a coverslip and

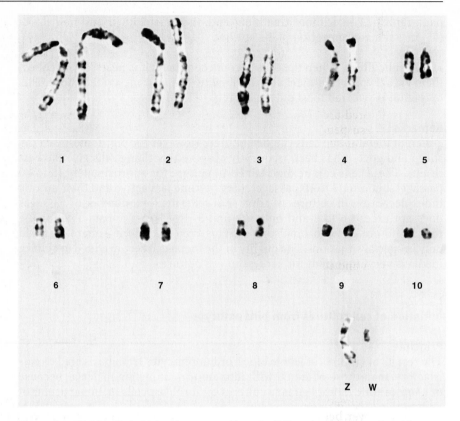

Figure I.1. RBG banded chicken karyotype (from Ladjali et al., 1995, reproduced by permission of the American Genetic Association).

the cells are counted under the microscope with a 10× objective in phase contrast. The number of the cells per ml in the cell suspension is N = the average number of cells per grid compartment multiplied by 10^5. To count white cells, the suspension can be stained using acetic violet, the red cells being lysed in this case (appendix M2). The cell suspensions can also be stained with trypan blue, which distinguishes non stained living cells, from dead cells, which are blue (appendix M3).

I.1.2.4
Cell storage

Principle
Cells cultured from tissue fragments can be stored at low temperature for many years without loss of viability if certain conditions are respected in order to minimise their alteration. Alterations of cell structures are due in particular to the effects of osmotic imbalances and to the formation of ice

crystals during freezing. Cells are therefore frozen and stored in liquid nitrogen ($-196\,°C$) in the presence of a cryoprotecting agent such as glycerol or dimethylsulphoxide (DMSO) and rates of freezing and thawing are carefully controlled.

Protocol

Cells to be stored are collected in exponential growth under standard conditions and resuspended at a final concentration of 10^6–10^8/ml in foetal calf serum containing 10% DMSO (appendix V1). 1 ml samples of the cell suspension are dispensed into plastic cryoconservation tubes which are placed in a polystyrene box and kept for about 24 hours in a freezer at $-80\,°C$, before transfer to a liquid nitrogen container.

For thawing, the tubes are removed from the liquid nitrogen container and rapidly placed in a water bath at $37\,°C$. As soon as the cell suspensions have thawed, they are diluted with MEM-10% FCS medium (appendix CM2), and centrifuged for 5 minutes at $400\,g$. The pelleted cells are resuspended in the same medium and inoculated into $75\,cm^2$ flasks also containing MEM-10% FCS medium.

I.2
Preparation of chromosomes in prophase or prometaphase

H. Hayes and B. Dutrillaux

Chromosome spreads are prepared from cells harvested during the phases of mitosis. However, between early prophase and late metaphase, chromosomes become more and more compact and short and, consequently the number of structures or bands revealed by chromosome banding techniques decreases. In many cases, chromosome identification is easier and more accurate, the larger the number of bands obtained and the better the resolution of their banding pattern. During the preparation of chromosome spreads by classical methods, the chromosomes are in different stages of prophase and metaphase, the majority being in mid or late metaphase because of the use of colchicine which blocks cell division at this phase. No method is available to accumulate cells before metaphase but it is possible to synchronise the divisions of a cell culture and to harvest them at a particular stage by precise monitoring of the cell cycle. Synchronisation methods exploit the property of certain chemicals to block the synthesis of one of the desoxyribonucleotide triphosphates necessary for DNA replication without interfering significantly with the synthesis of RNA and proteins. After release of the mitotic block, DNA synthesis resumes and the cells progress in a relatively synchronous manner through the various phases leading to mitosis. Populations containing large numbers of cells in prophase or prometaphase can thus be collected, according to the time of harvesting. The elongated chromosomes present in spreads prepared from such cells yield high resolution patterns of stained bands. For cytogenetic applications, the following synchronising agents are used: amethopterin or methotrexate, aminopterin, 5-fluorodeoxyuridine (FdU) and thymidine in excess.

I.2.1
Single or double synchronisation using thymidine

Lymphocyte cultures respond very well to synchronising agents and generally, a single treatment is sufficient. In contrast, for most other cultures (eg fibroblasts) two cycles of treatment are often necessary.

Principle

Thymidine, added to the culture medium, penetrates the cells and is quickly converted to deoxythymidine triphosphate (dTTP) (Figure I.2), the intracellular concentration of which depends on that of the extracellular thymidine. dTTP is an allosteric inhibitor of ribonucleotide reductase, the enzyme which reduces cytidine diphosphate (CDP) and uridine diphosphate (UDP) to dCDP and dUDP (Figure I.2). Hence, the addition of excess thymidine resulting in an excess of dTTP, blocks DNA synthesis by reducing the amount of dCTP synthesised. The block can be released by removing the excess thymidine either by repeated washing of cultured cells or by addition of dCTP to the culture medium. To achieve almost complete inhibition of DNA synthesis, the extracellular concentration of thymidine must be higher than 1 mM (Xeros, 1962) and a concentration of 3 to 5 mM is usually used. An important advantage of excess thymidine compared to other synchronising agents is its low toxicity and the fact that it can be used for both single and double synchronisation treatments. Double synchronisation which consists in the application of two successive blocks with excess thymidine, increases the number of cells synchronised at a given stage, because cells which escape the first block are synchronized during the second. The protocol described here (Hayes et al., 1991) is adapted from the method used in the laboratory of Dutrillaux and published by Viegas-Péquinot and Dutrillaux (1978).

Protocol for lymphocytes (single synchronisation)

Thymidine at a final concentration of 0.3 mg/ml is added to a culture of lymphocytes after 48–72 hours of growth [addition of 0.3 ml of sterile 1× thymidine solution, (appendix V3) to 10 ml of culture medium]. After one night (about 15 hours) at 37 °C, the tubes are centrifuged for 5 minutes at 400 g, the supernatant containing thymidine is removed, and the cells are suspended in BSS$^+$ (appendix B2) by pipetting, centrifuged again and washed once more. Finally, the cells are resuspended in growth medium (appendix CM1) containing 10 or 20% foetal calf serum, incubated for 5 to 7 hours at 37 °C and harvested according to the protocol described in paragraph I.4.

Protocol for fibroblasts (double synchronisation)

■ **Day 1.** Cells harvested after trypsin treatment are inoculated in 75 cm^3 flasks containing 15 ml of medium MEM-10% FCS (appendix CM2) at a

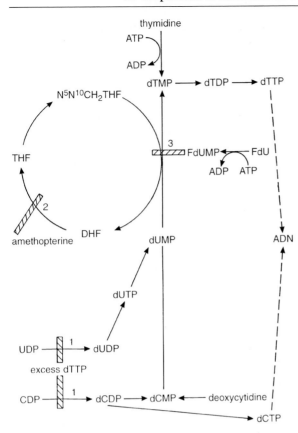

thymidine

Figure I.2. Different inhibition sites indicated by slashed bars of DNA synthesis used to synchronise cell cultures.

density between 1/5 and 1/10 of that attained at confluence and incubated at 37 °C for 3–4 hours. The concentration of thymidine is then brought to 3 mM by addition of 1.2 ml of sterile 1× thymidine solution (appendix V3) and the flasks are incubated for 14–15 hours, for the first block.

■ **Day 2.** Cells are washed three times for 15 minutes at 37 °C, with BSS⁺ buffer (appendix B2). They are then incubated again with 15 ml of MEM-10% FCS medium and after 8–9 hours, 1.2 ml of sterile 1× thymidine solution (appendix V3) is added for the second block during 14–15 hours. If only one thymidine block is applied, the cells are treated directly according to the protocol described in the next paragraph.

■ **Day 3.** Cells are washed again three times for 15 minutes with BSS⁺ buffer at 37 °C, and incubated again in 15 ml of medium MEM-5% FCS until the number of the dividing round cells observed under the inverted microscope, becomes constant. The cells are then treated with colchicine at a final concentration of 0.04 µg/ml (appendix V2) for 20 minutes at 37 °C. The further treatments (harvesting, hypotonic shock, fixation and spreading) are the same as described in paragraph I.4.

Remarks
This method does not synchronise all the cells in a culture, but it considerably increases the number of cells harvested in the required phase and makes possible the preparation of slides rich in prometaphase chromosome spreads. The time necessary for cells to reach early mitosis after release of the block varies according to the origin of the cells, the condition of the cultures and the presence or not of BrdU, (for the production of bands depending on BrdU incorporation, see paragraph II.3.1) but varies also from one culture to another. This time interval is usually about 6 to 8 hours but it may be longer. It is therefore necessary to carry out several tests to define the optimal conditions and it is advisable to prepare 3 flasks of culture in parallel for each experiment and to harvest one flask every 20 minutes as soon as the number of dividing cells begins to increase. Colchicine is added for 20 minutes before harvesting the cells to improve the dispersion of chromosomes during the preparation of spreads by inhibiting the formation of the mitotic spindle, and not to accumulate mitotic cells.

I.2.2
Synchronisation by amethopterin

Principle
Amethopterin or methotrexate is an analogue of folic acid and a strong inhibitor of dihydrofolate reductase (Blackley, 1969). This enzyme catalyses the reduction of folic and dihydrofolic acid to tetrahydrofolic acid and the inhibition of its activity blocks the synthesis of N^5 N^{10} methylene tetrahydrofolate which is the cofactor of thymidylate synthetase (Figure I.2). The synthesis of thymidine monophosphate (TMP) and, thus, that of DNA are blocked, and cells accumulate in the middle of the S phase of the cell cycle. The block is released by changing the culture medium and/or adding thymidine which is converted to thymidine triphosphate by the action of thymidine kinase and thymidylate kinase. Only one block can be performed by this method because amethopterin is very toxic to the cells.

Protocol
■ **Day 1.** Cells harvested after trypsin treatment are inoculated in $75\,cm^2$ flasks containing 15 ml of MEM-10% FCS medium (appendix CM2) at a density of about 2/10 to 3/10 of that reached at confluence and are incubated at 37 °C. About 5–6 hours later, 150 µl of a sterile 1× amethopterin solution (appendix V4) are added to yield a final concentration of $10^{-7}\,M$ (Yunis, 1976) and incubation at 37 °C is continued for 16 hours.

■ **Day 2.** Cells are washed once at 37 °C for a few minutes, with BSS⁺ buffer (appendix B2) and then reincubated in MEM-5% FCS medium supplemented with thymidine at a final concentration of 3 µg/ml (appendix V3). The rest of this procedure is the same as that described in paragraph I.2.1.

Remarks
The time required for cells to reach early mitosis after release of the block (5–7 hours) is generally shorter than in the case of synchronisation with thymidine. This may be due to the fact that after release of the block, the reinitiation of metabolic processes may be more or less rapid according to the blocking agent used. This method also gives more homogeneous populations of harvested cells usually blocked in early metaphase, which shows that the block induced by amethopterin is more rigorous than that caused by excess thymidine. It constitutes an alternative for the synchronisation of cell cultures which respond poorly to synchronisation induced by thymidine, but generally, the chromosomes obtained are less elongated.

I.2.3
Synchronisation by 5-fluorodeoxyuridine

P. POPESCU and B. DUTRILLAUX

Principle
As indicated in Figure I.2, a third blocking point exists during the reactions of endogenous synthesis of thymidine. This is the reaction catalysed by thymidylate synthetase which is inhibited by 5-fluorodeoxyuridine or FdU. Added at a concentration of 4 µM (appendix V6), FdU synchronises lymphocyte cultures by a block at the middle of the S phase and the results are similar to those obtained with amethopterin (Yunis, 1976) or an excess of thymidine (Viegas-Péquignot and Dutrillaux, 1978). The block can be released by the addition of thymidine or BrdU at a final concentration of 5 µg/ml (appendix V3 and V5).

I.2.4
Synchronisation by 5-bromodeoxyuridine or BrdU

Principle
5-bromodeoxyuridine or BrdU, an analogue of thymidine is frequently used in many chromosome banding techniques. Added at a final concentration of 0.2 mg/ml, BrdU blocks the cell cycle like thymidine and it can therefore be used directly to synchronise the cells (Dutrillaux and Viegas-Péquinot, 1981). In this case it is incorporated in the early replicated fraction of DNA, that is in R bands (see paragraph II.3.1). In order to obtain high resolution G banding after synchronisation with BrdU, it is necessary to discard the medium containing BrdU and to replace it by a standard culture medium without BrdU, for 6 to 8 hours. Chromosomes are then stained either with acridine orange (see paragraph I.5.3), or by the FPG technique (see paragraph II.3.1.2).

I.3
Direct techniques and very short term development cultures

In certain special cases, chromosome preparations can be obtained from spontaneously dividing cells ie without establishment of a true *in vitro* culture, although the cells can be maintained in culture for only a very short time. These techniques are useful for the study of neoplasia and in particular, leukaemia, using bone marrow cells or even peripheral blood.

I.3.1
Study of the bone marrow

Protocol
0.5 ml of bone marrow is aspirated from the sternum or iliac crest by the use of a Mallarmé syringe. The inoculation of cultures is carried out under the same conditions as for the culture of lymphocytes, but without phyto-haemaglutinin. The aspirated marrow should be distributed immediately into 3 tubes, each being inoculated with 0.1 ml of marrow. Colcemid at a final concentration of 0.04 µg/ml (appendix V2) is added to one of the three tubes and its contents are directly treated, without incubation. The other two tubes are incubated at 37 °C for 24 and 48 hours respectively and then treated according to the standard protocol for chromosome preparation (see paragraph I.4).

Remarks
The results obtained vary according to the origin of the bone marrow, which explains the need for three cultures harvested at three different times. Dutrillaux and Couturier (1981) showed that cultures of human marrow harvested after 24 hours give the best results.

I.3.2
Study of bird chromosomes from feather pulp

The genetic complement of birds is characterised by a large number of chromosomes classified into two categories: macrochromosomes and microchromosomes, and by the heterogamety of the female.

In birds, there exist dimorphic species in which the male and the female can be distinguished by direct phenotypic observation and monomorphic species in which the male and the female are identical. In this case, bird breeders use several techniques for sex identification: endoscopy, assaying sex hormones in dung and chromosome analysis. Chromosomal sexing has the advantage that it can be carried out whatever the age of the bird, whereas endoscopy is possible only in birds at least 3–6 months old according to species, and sex hormone assays are applicable only in the case of mature birds. In addition, the latter method is aleatory in those species which present a seasonal loss of sexual activity.

Chromosomal analysis using feather pulp cells can be performed as soon as pin feathers are present at the base of growing feathers (Shoffner et al., 1967). Wing feathers usually yield more active cultures although any growing feather can be used. If none is found, mature feathers should be pulled to initiate growth of new ones which will be ready in 10–20 days.

Protocol
One or two 2 or 3 week old feathers are pulled after local disinfection. The feather extremity is disinfected with alcohol and the semi-solid pulp is extracted by pinching off 2–3 mm of the base with a forceps. The pulp is dilacerated and placed in a culture flask containing Ham F12, TC199, or RPMI1640 medium (appendix CM1) with 10 to 20% foetal calf serum and colchicine (appendix V2) at a final concentration of 10 µg/ml. The flask is incubated at 37 °C for 2 hours. The cell suspension is then transferred to a tube and centrifuged at 400 g for 5 minutes and the pellet is resuspended in 5 ml hypotonic solution of Na_3citrate $2H_2O$ (0.45% weight/vol.) and incubated at 37 °C for 15 minutes. The cells are then fixed in fresh fixative (3 vol. ethanol – 1 vol. acetic acid) for at least one hour. Chromosome spreads are then prepared as for mammalian cells.

Remarks
For bird sexing, some authors advise *in vivo* treatment with colchicine (by injection of 0.01 ml/7.5 g of body weight of a colchicine solution (10 µg/ml appendix V2), either intraperitoneally or into the wing vein of the bird). From 1 to $1^1/_2$ hours later, feather pulp is collected and cells extracted from it are incubated directly in the hypotonic solution without culture.

Bird sex can be determined by chromosome analysis in most species. If the karyotype of the analysed species is known, the sex chromosomes can be identified in chromosome spreads stained by the classical Giemsa technique (see paragraph I.5.1). If not, heterochromatin staining (CBG bands, see paragraph II.5.1) permits unequivocal identification of chromosome W which appears as a dark spot in chromosomal spreads at prophase and metaphase of most species of birds (Figure I.3) even with mediocre chromosome preparations (Stefos et Arrighi, 1971). Bird sexing by chromosome analysis is limited by the fact that it requires special competence and is expensive.

I.3.3
Study of fish chromosomes

Chromosome analysis in fish is difficult because of the small size and large number of chromosomes. For this reason, it has been limited for several decades, to the simple determination of the chromosome number which is used as a taxonomic criterium in many species.

Two protocols for lymphocyte culture, one from blood and the other from kidney or spleen and a protocol for chromosome banding by BrdU incorporation in live fish are described in this paragraph.

Figure I.3. CBG banded chicken chromosomes obtained from feather pulp (photograph P. Popescu). Chromosome W appears as a dark spot.

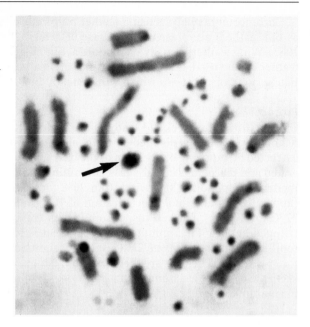

Protocol for culture of blood lymphocytes

0.5 ml of blood is collected and centrifuged at 400 g for 5 minutes. A sample of 0.25 ml of plasma rich in lymphocytes is taken from the supernatant and inoculated into 5 ml of TC199 medium (appendix CM1) containing phytohaemagglutinin P at a final concentration of 50 µg/ml and 10% foetal calf serum. The tube is incubated at 20 °C for 5 days and chromosome spreads are then prepared according to the usual protocols. The PHA P can be replaced by LPS (lipolysaccharide), at a concentration of 100 to 500 µg/ml, which produces a better mitotic index in most cases (Hartley and Horne, 1985).

Protocol for culture of lymphocytes from kidney or spleen (Medrano et al., 1988)

The live fish is kept for a few hours in an aquarium in water containing 10 U.I./ml penicillin and 10 µg/ml streptomycin. The kidneys and the spleen are then removed under sterile conditions, washed in sterile PBS⁻ solution (appendix B1) and dilacerated by passage through a sterile stainless steel grid (pore diameter about 0.5 mm). The mixture of lymphocytes and red cells thus obtained is then washed several times in sterile PBS⁻ solution, the number of cells is determined (see paragraph I.1.2.3) and they are inoculated at a concentration of 10^7–5.10^7/10 ml into TC199 medium (appendix CM1) containing 20% foetal calf serum, 50 µg/ml phytohaemagglutinin P, 100 UI/ml penicillin and 100 µg/ml streptomycin. The cells are incubated for 5–6 days at 18 °C, or 24 °C for tropical species. Cell harvesting and the chromosome preparation are carried out according to the usual protocols.

Protocol for chromosomal banding by BrdU incorporation in live fish

A technique for chromosomal banding by direct incorporation of BrdU in live fish has been described by Almeida Toledo and colleagues, (1988) and produces banding patterns of sufficient quality to permit comparison of the karyotypes of different species and the search for chromosome homoeologies. In this method, 1 ml/100 g body weight of a solution containing 0.1 mg/ml BrdU and 0.02 mg/ml FdU [326 μl of 1× BrdU working solution (appendix V5) + 2.7 ml of 1× FdU working solution (appendix V6) + PBS⁻ solution (appendix B1) for 10 ml], is injected into the fish before chromosome preparation from kidney or spleen. The fish is kept in an aquarium either for a short treatment (5–8 hours, incorporation during late S and early G2 phases), or for a long treatment (14 hours, incorporation during the whole S phase or part of the S phase and the whole G2 phase). Chromosome preparations are then stained either with acridine orange (see paragraph I.5.3), or by the FPG technique (see paragraph II.3.1.2) or with quinacrine (see paragraph II.2.1) as shown in Figure I.4.

Remarks

Stimulation of fish lymphocytes using PHA or LPS does not always produce a positive result and in some fishes it is always ineffective (Hartley and Horne, 1985).

Medrano and colleagues (1988) have also used a technique of direct culture of cells obtained from kidney, spleen or epithelial tissue, after an *in vivo* colchicine treatment but the quality of the chromosome banding is less good.

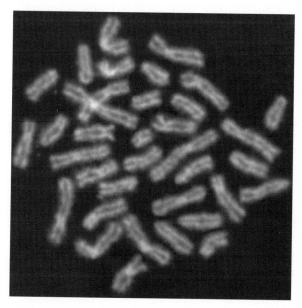

Figure I.4. Fish chromosomes stained with quinacrine (*Anguilla anguilla*) (photograph B. Dutrillaux).

I.3.4
Study of insect chromosomes

In general, cytogenetics owes much to the pioneering studies on *Drosophila* by researchers such as Morgan, performed at the beginning of this century. However, most recent progress has been made with mammalian and, particularly, human material. This relative decline in the amount of data obtained from studies with insects is due to the fact that, except for Dipterous insects and other species such as Collembols in which giant chromosomes (see chapter VI) can be observed in somatic cells, it is often difficult to obtain cells which give good chromosome spreads. Although, such cells do exist in many tissues, it is very difficult to determine the best time to carry out chromosome preparations. This time varies greatly, within the same order, from one family to another, even from one genus to another. In addition, cell development cycles can be interrupted by diapauses, during which all the cells are quiescent.

Considering the diversity of species and situations it is not possible to describe, case by case, the best procedure. A few examples are presented in the next paragraph.

I.3.4.1
Preparation of metaphase spreads from embryo cells

Embryo containing eggs are a good material source provided that their origin can be precisely determined. Therefore, this method is used mainly when insect breeding installations are accessible.

Protocol
The eggs washed externally with physiological serum (NaCl 0.9% weight/vol.) are placed in a conical tube containing culture medium (appendix CM1) and colchicine at a final concentration of $0.04\,\mu g/ml$ (appendix V2). The eggs are then crushed with a well fitting piston. Debris are removed by pipetting, taking care to retain only the cellular material released from the eggs. This is suspended by repeated pipetting and the suspension is left for 2–3 hours at room temperature and then centrifuged for 5 minutes at $500\,g$. The supernatant is removed by pipetting and the cell pellet is resuspended in a hypotonic solution (1 vol. animal serum – 2 vol. distilled water), then left at room temperature for 10 minutes. After a second centrifugation for 5 minutes at $400\,g$, cells are fixed and spreads prepared as for mammalian cells.

I.3.4.2
Preparation of metaphase spreads from gonads

Division figures can also be obtained from spermatogonia and ovogonia. It is however difficult to determine the period of active gametogenesis which can occur either in the larval stages or in the imago. According to our experience with Coleoptera it is necessary to work on the latest larval stage in

Cerambycidae, whereas in Scarabaeidae, Carabidae and Chrysomelidae imagos still give results. However, as mentioned above, it is better to do preliminary tests species by species.

Protocol

Male gonads are more easily observed than female gonads under the binocular microscope and in addition, the testes are richer in dividing cells. After location the gonads are placed in a watch glass and dilacerated as finely as possible in a few drops of physiological serum (NaCl 0.9% weight/vol.). This suspension is transferred to a conical tube and culture medium (appendix CM1) containing colchicine at a final concentration of 0.04 µg/ml (appendix V2) is added. Further steps in this protocol are the same as those described for preparations from embryonic cells.

I.3.4.3
Chromosome analysis from other tissues

Besides the squash methods frequently used for polytene chromosomes which are mainly observed in the salivary glands of Diptera (see chapter VI), other methods are used for the preparation of somatic chromosomes. In particular, in Drosophila, good quality chromosome spreads are obtained from cerebral ganglions of third stage larvae (see paragraph VI.1.2). The technique described above for the embryonic cells can be used without difficulty.

I.3.4.4
General comments

The main difficulty in work on insects, once the problem of obtaining cells has been solved, is the small number of cells available. We have performed many experiments on slides in a single drop which limits the number of lost cells. However, spreads obtained by this method are rarely as good as those produced by the technique involving centrifugation. For this reason, we have finally adopted the latter. Knowing that excessive purification of the population of mitotic cells is not advantageous, two strategies can be used to overcome the problem of the small quantity of material available. One is to combine somatic tissue of another origin with the selected tissue. This has the disadvantage of reducing the mitotic index but permits analysis with material from a single animal. The other is to treat the cells of several animals together, which is satisfactory in determining the karyotype of a species but it cannot be used to analyse individual variations.

The techniques described here can also be used to study chromosomes in meiotic stages, the major difficulty being again the determination of the best time to operate, which varies from one species to another.

I.4
Preparation of chromosome spreads

H. HAYES

To obtain chromosome preparations, mitotic cells must be harvested and then incubated in a hypotonic solution to produce good chromosome dispersion, fixed and spread on slides.

Principle
■ **Accumulation of mitotic cells.** The population of a cell culture at any given time contains a mixture of cells at all stages of the cell cycle. For chromosome analysis, this population must be enriched in mitotic cells. This is achieved by using drugs (colchicine, colcemid) which block chromosome division at anaphase by inhibiting the formation of the mitotic spindle fibres to which chromosomes attach via their centromeres before dividing. In cell cultures treated with these drugs for 1–2 hours, an accumulation of cells in metaphase is observed.

■ **Hypotonic treatment.** To facilitate cell disruption and the dispersion of chromosomes when the cell contents are spread on slides, the cell suspension is incubated in a hypotonic solution which causes osmotic swelling of the cells. Different hypotonic solutions are used: diluted salt solutions, KCl 75 mM, and diluted solutions of serum or plasma. Some techniques use in addition EDTA which by chelating divalent cations destabilises and fragilises cell membranes.

■ **Fixation.** Cell suspensions are fixed to preserve the internal structure of the cells and allow use of chromosome staining and banding methods. Chromatin fixatives generally contain acetic acid.

Protocol
■ **Metaphase accumulation.** A solution of 1× colchicine (appendix V2) is added to the cell culture at a final concentration of 0.04 μg/ml, 90 minutes before harvesting the cells and incubation is continued at 37 °C.

■ **Hypotonic treatment.** The cell suspension is centrifuged for 5 minutes at 400 g, either directly, in case of lymphocyte cultures or after trypsinisation in the case of fibroblast cultures. The pellet is gently resuspended in 9 ml (for 25 cm² culture flask) or 18 ml (for 75 cm² culture flask) of a hypotonic solution consisting either of 1 volume of foetal or newborn calf serum and 5 volumes of distilled water supplemented with 4 μg/ml of EDTA, or of 75 mM KCl (appendix M1) which has been previously filtered through a 0.45 μm membrane and warmed to 37 °C. The cell suspension is then incubated in a waterbath at 37 °C for 13 minutes.

■ **Fixation.** A prefixation is effected by adding 1 ml of fixative (3 vol. ethanol – 1 vol. acetic acid for lymphocytes or 3 vol. methanol – 1 vol. acetic acid for fibroblasts) directly to the hypotonic medium, for 5 minutes. The cell suspension is then centrifuged for 5 minutes at 150 g, and the pellet is dispersed in 0.5 ml of fixative to obtain a good homogenisation of the suspension. A further 6 ml of fixative is then added and the suspension is held at room temperature for 20 minutes. After a second centrifugation for 10 minutes at 400 g, the pellet is dispersed in 6 ml of fixative and the tubes are kept at 4 °C for at least 15 minutes, but can be left at 4 °C until the next day before continuing the preparation.

■ **Spreading.** After fixation and centrifugation, the cells are resuspended in the minimum volume of fixative. After observation of the first slide, this volume is adjusted to obtain a correct spreading density. The cell suspension can be spread on either wet or dry slides.

Protocol for wet slides
The cell suspension is dropped onto slides which have been previously washed, according to the protocol in appendix M5 and kept in cold distilled water, by permitting one drop to fall from a height of about 10 cm onto each cold slide covered by a thin water film. Water in excess around the drop is absorbed with tissue paper (important for further analysis) and the slides are left to dry at room temperature. They are then frozen at –20 °C and can be stored for several years.

Protocol for dry slides
Clean and dry slides (type Esco) are placed on wet filter paper. The cell suspension is spread by dropping one drop on each slide from a height of about 10 cm. The slides dry very quickly. Since the drops spread on a smaller surface than in the previous protocol, the cell suspension must also be more diluted which permits preparation of a larger number of slides. This type of spreading is very suitable for *in situ* hybridisation because less halo is observed around the chromosomes revealed by counterstaining. The slides can be kept by freezing, as above.

Remarks
In the case of fibroblast cultures a more selective harvesting of the mitotic cells is possible. This can be achieved by reducing the duration of trypsinisation or even, depending on the cell preparation, eliminating this treatment and shaking the culture flasks in order to detach the mitotic cells which are roundshaped at this stage. Cells from 5 to 10 culture flasks must then be combined, to obtain a cell pellet large enough to permit the further stages of chromosome preparation.

Chromosome preparations obtained according to the protocol using six fold diluted serum are generally of better quality (less swollen) compared with those obtained using 75 mM KCl. Although chromosomes obtained in this way are less well dispersed, this facilitates their use for the analysis of banded karyotypes and also for *in situ* hybridisation.

I.5
Staining techniques of chromosome spreads

H. HAYES and B. DUTRILLAUX

I.5.1
Classical staining using Giemsa

This is the most frequently used method for staining and normal or phase contrast microscopy of the chromosome bands produced by different banding techniques.

Principle
Giemsa stain is a complex mixture of positively charged molecules mainly from the thiazine group, which interact with the phosphate groups of DNA. The stain changes colour from light to dark blue (Comings, 1978) because binding of the thiazines to DNA reduces the wavelength of the stain's absorption peak.

Protocol
The staining solution is prepared as follows immediately before use:

- Sorensen's buffer solution pH 6.8 (appendix B3):   3 ml
- Giemsa rapid R stain solution (R.A.L. Prolabo, Paris):   3 ml
- distilled water: 94 ml

The slides are incubated for 10 minutes in the staining solution at room temperature, rinsed thoroughly first in running tap water and then in distilled water and finally air dried.

Remarks
The Giemsa R stain is usually used at a concentration of 3% but this can be adjusted from 1 to 5% according to the intensity of the observed staining. The banded preparations are observed under the microscope, either in normal light or in phase contrast. Photographs are taken using a Kodak technical pan film at the sensitivity which is optimal for the type of the microscope used.

I.5.2
Orcein staining

Orcein is not a chromatin specific stain, but in acid solution it acts like a basic stain and is fixed on the basophilic chromosome which it stains dark purple red (Figure I.5). For this reason, aceto-orcein stain is used instead of orcein

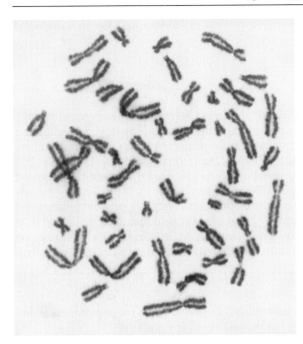

Figure I.5. Human chromosomes stained with acetic orcein (photograph B. Dutrillaux).

alone. Here, we describe two protocols for the preparation and use of this stain.

Protocol 1

The staining solution is prepared by heating a mixture of pure acetic acid (45 ml) and orcein (1–2 g, Gurr) to dissolve the orcein. The solution is cooled, diluted with 55 ml of distilled water and filtered through Whatman N°1 paper before use.

To stain the preparation, a few drops of stain are deposited onto the slide and after 3–5 minutes it is rinsed in distilled water and dried. A coverslip is then mounted and luted and the preparation is observed in normal light.

Protocol 2

The staining solution is prepared by heating a mixture of orcein 2 g (Gurr) and 33 ml of acetic acid, to dissolve the orcein. 33 ml of lactic acid and 33 ml of water are then added and while the solution is still warm, it is filtered through Whatman N°1 paper. The product can be stored for several months at 4 °C requiring only centrifugation and filtration from time to time.

Remarks

This method was extensively used in the past but has been progressively abandoned during the 1960's and especially in the 1970's when the use of the

Giemsa stain was introduced. However it remains in use for staining polytene chromosomes (see chapter VI).

I.5.3
Acridine orange staining

Acridine orange became very popular in cytogenetic studies especially in the 1970's. Besides its aesthetic interest, arising from its property of producing a double, red and green, chromosomal stain under certain conditions, it also has the property of enhancing R banding (Dutrillaux et al., 1973). For this reason, it is still used in some laboratories (Figure I.6). In addition, it is the only method which gives good T banding.

Acridine orange interacts differently with single and double stranded nucleic acids. It can polymerise on the single stranded nucleic acids and then emits a dark reddish fluorescence. In contrast, it is fixed as a monomer on double stranded DNA and emits an intensive green fluorescence under ultraviolet illumination.

Protocol
The staining method is very simple but the subsequent rinsing process is delicate and is very important for the quality of the staining.

The staining solution is prepared by mixing 95 ml of Sorensen's buffer pH 6.8 (appendix B3) and 5 ml of acridine orange stock solution (1 mg/ml, appendix S8). This solution can be stored for at most one week at room temperature.

The slides are incubated for 20 minutes in this solution, rinsed immediately in running tap water for 1–2 seconds, then rinsed several times by pipetting Sorensen's buffer (appendix B3). They are mounted with a coverslip in Sorensen's buffer and excess buffer is removed with blotting paper. They are examined by fluorescence microscopy using a filter block appropriate for acridine orange. The quality of the rinsing steps is evaluated by microscopic examination: too greenish a result indicates too vigorous rinsing and too reddish a result insufficient rinsing. Staining or rinsing can be repeated according to the result observed.

Remarks
Whatever R-banding treatment is used prior to staining (BrdU incorporation or denaturation) it should be noted that the longer the slides have been stored (except for storage at −20 °C) the more the red fluorescence predominates and the more thorough the rinsing process should be.

If denaturation is used, the intensity of the red staining and the number of structures having this colour increase according to the intensity of the treatment. Thus, too greenish staining indicates insufficient denaturation. In this case, the same slide can be treated again. In contrast, too reddish staining indicates excessive denaturation which has left only denatured single stranded DNA. In this case, another slide must be taken.

It must also be borne in mind that staining is modified during illumination of a slide in the microscope field. At first, during a period which corre-

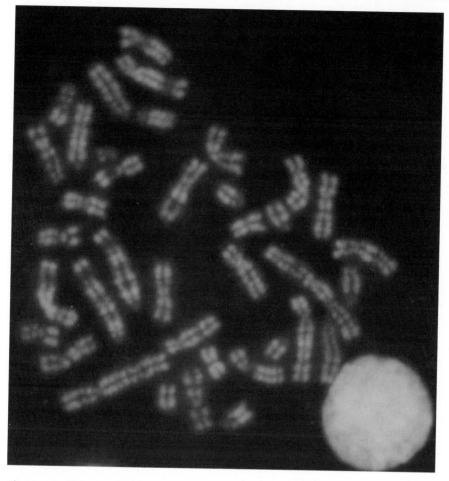

Figure I.6. RBA banded domestic pig chromosomes stained with acridine orange (photograph P. Popescu).

sponds approximately to the time required to complete its examination, the reddish staining decreases relatively to the greenish staining probably because the autopolymerised stain molecules which emit a red fluorescence are depolymerised and detached by ultraviolet light. Since this red fluorescence masks the green fluorescence of underlying regions of the chromosome, the latter appears progressively more and more greenish, except for the regions containing only denatured DNA, which remain dark reddish. During prolonged ultraviolet illumination, the green fluorescence also decreases progressively. However, observation under optimal conditions is possible during a period varying from several minutes to more than one hour.

Chromosome staining is always more intensive after BrdU treatment than after denaturation treatment. The red stain indicates regions of BrdU incorporation or denaturation according to the banding method used.

I.5.4
Propidium iodide or DAPI staining

These stains are used for the simultaneous examination in fluorescence microscopy of the chromosome bands induced by BrdU incorporation and the hybridisation signal on chromosomes treated for *in situ* hybridisation. These techniques are described in paragraph II.3.1.3 for propidium iodide and paragraph II.3.1.4 for DAPI.

II
Chromosome Banding Techniques

II.1
Introduction

H. Hayes

The most extensively exploited and paradoxically the least well understood property of chromosomes is that of displaying along their arms a succession of more or less strongly coloured bands after staining or other special treatments. These reproducible and distinctive banding patterns are characteristic to each chromosome. They permit the identification of chromosomes, the construction of karyotypes, detailed description of chromosomal or structural rearrangements, localisation of markers and the comparison of chromosomes within a species or between closely or distantly related species. However, the understanding of chromosome banding mechanisms is far from complete and the explanations proposed in the literature are hypothetical. This chapter describes the types of structures and more exactly, the types of bands which can be observed in chromosomes and their known characteristics.

II.1.1
Chromosome organisation

The chromosomal genetic material of most eukaryotic cells consists of linear DNA molecules (one molecule per chromosome). Each of these molecules contains several million nucleotides; the combined sequences of all these molecules determine the genetic information present in each organism. In the cell nucleus, DNA is generally associated with other molecules, mainly histones, small basic proteins present in about the same quantity as DNA and to a lesser extent with acid nonhistone proteins which represent 10–30% of the total. This DNA-protein complex also called chromatin because it fixes coloring agents, constitutes the chromosome structure. The histones which are essential for the folding of chromosomal DNA play a central role in chromosome structure. Chromosomal DNA molecules associated with proteins and RNA pass through different degrees of compaction during the various stages of the cell cycle and it is only at the stage of maximum compaction during mitosis that they can be observed as distinct entities in the form of chromosomes. However, it should be noted that since the introduction of total genomic or chromosome specific DNA as probes for *in situ* hybridisation (Manuelidis, 1985; Schardin et al., 1985; Lichter et al., 1988a; Pinkel et al., 1988) it is possible to localise the region occupied by an individual chromosome in the interphase nucleus.

Several types of structures can be observed according to the treatments and staining procedures used for chromosome examination (Comings, 1978; Sumner, 1989):

– **euchromatin** revealed by chromosome staining methods which produce bands denominated Q, G, and R corresponding to transcribed DNA. The names of these bands Q for quinacrine, G for Giemsa and R reverse are derived from the treatments and staining techniques used to produce them, but distinctive and superposable band patterns are produced in each case (fluorescent Q^+ bands = strongly stained G^+ bands = weakly stained R^- bands / weakly fluorescent Q^- bands = weakly stained G^- bands = strongly stained R^+ bands). R (reverse) bands discovered after Q bands were called reverse because the pattern of strongly stained R bands is the same as those of weakly stained Q and G bands and vice versa. Conventionally, the terms Q, G and R bands designate positive, strongly stained bands. The characteristics of euchromatin will be described later.

– **constitutive heterochromatin** intensely and selectively stained by C banding techniques (C from centromere). Constitutive heterochromatin the function of which is unknown at present is a fraction of chromatin which does not decondense during interphase and which is located in the juxtacentromeric regions of all chromosomes in almost all species, but sometimes also in other regions. It contains highly repetitive DNA sequences rich in AT or GC base pairs (according to chromosomal location and species) and is replicated very late during the S phase of DNA synthesis.

– **telomeric regions** are revealed by T banding (T from terminal). T bands are generally considered to be composed of euchromatin because they constitute a subfraction of R bands, particularly resistant to thermal denaturation because of their high GC content. Telomeres are complex structures situated at the ends of chromosomes and are involved in the stability of chromosomes and the replication of their DNA. They contain tandem repetitions of short GC rich sequences. Their sequences are known and they can be detected by *in situ* hybridisation. Telomeres become shorter as cells age ie parallel with the number of cell divisions (Broccoli and Cooke, 1993), which explains their instability.

– **centromeres** appear as a constriction on each chromosome after Giemsa staining. It can be shown by immunofluorescent techniques that these constrictions also contain specialized structures named kinetochores by means of which chromosomes attach to the meiotic and mitotic spindles. The kinetochores are considered to form part of the constitutive heterochromatin. In primates (Willard and Waye, 1987), including man, the centromere regions are rich in repetitive alpha satellite or alphoid DNA (DNA consisting of a series of tandem repeats of a 171 base pair sequence) while adjacent regions contain classical satellite DNA. In other mammals, such as mouse, goat and cattle, centromeric DNA is also composed of tandemly repeated sequences, which appear to be specific for each species, but the organisation of which is not so well known.

- **nucleolar organiser regions** (NOR) stained by the Ag-NOR method. These structures participate in the formation and maintainance of the nucleolus in interphase nuclei.

II.1.2
Euchromatin

In mammals, the most interesting component of chromosomes for their identification is euchromatin. In birds, macrochromosomes can be identified by banding techniques which mark euchromatin, but microchromosomes cannot be identified by this technique, and *in situ* hybridisation using specific centromeric probes or known gene sequences must be used. In fish chromosomes, only certain banding techniques (eg BrdU incorporation, see paragraph I.3.3) produce bands in euchromatic regions. On the contrary, for mitotic chromosomes of some insects and amphibians the best results are obtained by using techniques which stain heterochromatic regions.

The euchromatin of all mammalian chromosomes is subdivided into two fractions containing nearly equal amounts of genomic DNA, the G and R bands. A series of opposed functional and biochemical characteristics are associated with each of these fractions. The most important of these characteristics are summarized in Table II.1 (Holmquist et al., 1982; Holmquist, 1989; Bickmore et Sumner, 1989; Burkholder, 1993; Gardiner, 1995).

According to this table, the chromosome bands can be considered either as DNA segments with specific contents of base pairs, genes or dispersed repetitive sequences, or as functional units with specific replication and condensation properties. Until the middle of the 1980's, it was considered that

Table II.1. ··

G bands	R bands
fluorescent Q bands	non fluorescent Q bands
negative R bands	positive R bands
chromomeric regions at the pachytene stage of meiosis	interchromomeric regions
late replication during the S phase	early replication
AT rich	GC rich
methylcytosine poor	methylcytosine rich
few genes	most housekeeping and tissue specific genes
SINE sequences	LINE sequences

the bands reflected a longitudinal differentiation of the nucleoprotein structure of the chromosomes but the nature of this differentiation was unknown.
The organisation of the mammalian genome has been the subject of many
studies which have led to different models for the division of the genome
into regions differing in base composition or in their content of dispersed
repeated sequences of the Alu type, for example the isochore model (Bernardi
et al., 1985) and the "flavor" model of metaphase chromatin (Holmquist,
1992). Only studies on the distribution of certain types of dispersed repetitive sequences (Manuelidis and Ward, 1984; Korenberg and Rykowski, 1988)
and on the relation between bands and the chromosome structure (Saitoh
and Laemmli, 1994), which opened new perspectives, will be described briefly
here. Recent revues by Burkholder (1993) and Gardiner (1995) can also be
consulted.

Distribution of SINE and LINE sequences along chromosomes
Dispersed repeated sequences fall into two categories:

- "Short Interspersed Repeated Sequences" or SINEs are sequences less than
  1 kb in length, of which 3.10^5 to 9.10^5 copies are distributed throughout the
  genome
- "Long Interspersed Repeated Sequences" LINEs are sequences 6 to 7 kb in
  length distributed throughout the genome in about 10^4 to 10^5 copies.

In man, to an extent of more than 90%, SINE sequences correspond to the
300 bp Alu sequences so named because they generally contain a restriction
site for the enzyme *AluI* and LINE sequences to the L1 or *KpnI* sequences so
named because they can be isolated in part after the digestion of genomic
DNA with the restriction enzyme *KpnI*. The localisation of these sequences
along the human chromosomes by *in situ* hybridisation has shown that they
are not randomly distributed (Manuelidis and Ward, 1984; Korenberg and
Rykowski, 1988). The Alu sequences are found predominantly in R bands
where they represent about 18% of the total DNA, whereas the L1 sequences
are found predominantly in G bands in which they represent about 14%
of the total DNA. These findings suggest that they could be implicated in at
least some opposed characteristics of these two fractions of euchromatin
(Korenberg and Rykowski, 1988; Bickmore and Sumner, 1989).

Differential arrangement of AT alignment in Q/G and R bands
The group led by Laemmli which has studied the structure of the protein
skeleton of the chromosome for the past ten years has recently proposed a
model based on a relation between the organization of the chromatin fibre
and the succession of bands along the metaphase chromosomes (Saitoh and
Laemmli, 1994). As mentioned above, chromatin passes through different
degrees of compaction according to the stage of the cell cycle. The model of
Saitoh and Laemmli proposes that during one of these degrees of compaction, the chromatin fibre is organised in loops, the bases of which are
bound to the protein skeleton via specific DNA regions named "SAR" (Scaffold Associated Regions, Gasser and Laemmli, 1987). These SAR regions

contain special nucleotide sequences, very rich in AT base pairs (>65%). Because of their role as anchors of the chromatin fibre to the chromosome skeleton, these SAR regions must necessarily be arranged non randomly and constitute a central structure along the axis of the chromatin fibre (Saitoh and Laemmli, 1994). Using confocal microscopy, these authors examined the structure of chromosomes stained by daunomycin, a fluorochrome specific for DNA sequences containing more than 65% AT base pairs, and on the basis of their results suggest that the sequence of stained and non stained bands revealed by daunomycin is the result of a non uniform distribution of the alignment of the most AT rich sequences and therefore of the alignment of AT rich regions along the chromosome. They propose that this alignment is denser in Q bands due to spiralisation of the chromosome skeleton with which it is associated whereas in R bands it remains in a linear and therefore less condensed form. Thus, the banding pattern observed by staining AT rich motifs with daunomycin is due not to variations in the overall base composition of chromatin loops but to different densities of the alignment of the very AT rich sequences which they contain caused by different degrees of compaction of the chromatin in Q and R bands.

These results lead to a better understanding of genome organisation and the mechanisms of chromosome banding. It must be emphasized here, that in addition to the Q, G and R bands, referred to as "structural" bands because they depend only on the intrinsic nature of the chromosomes, so called "dynamic" bands are produced by the incorporation of a modified base into the DNA of the living cell and therefore depend on replication. The banding patterns observed are the same as those of structural G and R bands.

II.1.2.1
"Structural" bands

Q bands

Q banding patterns produced by quinacrine or by 4'-6-diamidino-2-phenylindole (DAPI) are the result of differential excitation and extinction of the fluorochrome according to the AT content of DNA and to the presence or absence of certain proteins. Weisblum and de Haseth (1972) for quinacrine and Lin and his colleagues (1977) for DAPI have shown that these fluorochromes bind preferentially to AT base pairs and that this binding intensifies their fluorescence as compared to binding to GC base pairs.

Knowing that the base pair compositions of LINE L1 sequences (58% AT) and SINE Alu sequences (44% AT) is correlated with those of the AT rich Q or G bands and of the GC rich R bands this could be one of the factors involved in the production of the Q banding pattern by fluorochromes such as quinacrine and DAPI. However, the differences between the overall AT or GC base compositions of Q and R bands are not sufficient to explain the production of Q bands using daunomycin because this fluorochrome is specific for DNA containing >65% AT base pairs. This emphasizes the interest of the model proposed by Saitoh and Laemmli (1994).

Q bands are generally weakly stained and must be observed under fluorescence microscopy. Therefore, they are no longer used routinely in laboratories studying human genetics. Nevertheless, since their production requires no special treatment, these bands can be used as complement to other methods, particularly for *in situ* hybridisation.

G bands

The G banding pattern of metaphase chromosomes corresponds to that of the chromomeres observed without treatment in chromosomes at the pachytene stage of meiosis (Ferguson-Smith and Page, 1973; Okada and Comings, 1974; Luciani et al., 1975). It is therefore interesting that G banding methods use either the effect of a proteolytic enzyme or mild denaturation conditions ie treatments which affect the interactions that stabilise the structures of the different protein and nucleic acid components of the chromatin. Several hypotheses have been proposed to explain the mechanism of G banding:

- Dutrillaux and colleagues (1972) and Bostock and Sumner (1978) suggest that this mechanism is based mainly on differences in protein composition along the chromosome, whereas
- Comings (1978) believes that G banding techniques cause chromatin rearrangement leading to exposure of the band pattern of meiotic chromomeres.

Support for the first hypothesis may be obtained by studies of the distribution of dispersed repeated sequences. Since, in man, R bands are the sites of Alu sequences and G bands of L1 sequences and since each of these sequences has a characteristic base composition. Research is now in progress to identify the proteins which interact with each of these two types of sequences and to investigate their biochemistry.

Saitoh and Laemmli developed the second hypothesis when they observed that the chromosome skeleton was stained more intensively by Giemsa after partial decondensation. They suggest that the treatments used in G banding such as trypsin treatment, lead to the unfolding of the chromatin loops and permit the Giemsa staining of the protein skeleton associated with the alignment of AT rich sequences.

R bands

R bands correspond to GC rich euchromatin regions and it seems that as for Q bands base composition is one of the main factors involved in the production of these bands. This hypothesis is compatible with the following observations:

- treatments (Comings, 1973; Dutrillaux and Covic, 1974) which systematically produce R type banding, involve denaturation followed by differential renaturation of DNA and it is known that the richer DNA is in GC base pairs the more resistant it is to denaturation. These treatments thus permit the GC rich regions of the denatured DNA (R bands) but not the AT rich regions (G bands) to renature.

- R banding patterns can be obtained without denaturation, by the use of fluorescent compounds such as chromomycin A3 (Schweizer, 1976), olivomycin (Van de Sande et al., 1977) and mithramycin (Schnedl et al., 1977) which have a strong affinity for GC base pairs.
- the R bands of human chromosomes contain Alu sequences which denature more slowly and renature more rapidly than the L1 sequences because they are more GC rich than the latter, shorter and therefore less complex, and often arranged in inverted repeats. Consequently, after denaturing treatments, only the DNA of R bands in which Alu sequences are predominant remains undenatured or renatures spontaneously whereas that of G bands in which L1 sequences are predominant is denatured.

II.1.2.2
"Dynamic" bands

Observation of dynamic bands (Figure II.1) involves incorporation of a nucleotide analogue (e.g. 5-bromo-2′-deoxyuridine or BrdU) into chromosomal DNA during a specific interval in the S phase of the cell cycle. The substitution of thymidine by BrdU modifies the staining characteristics of the chromatin. Since euchromatin contains two fractions, one of which replicates during the first half of the S phase (early replication) and the other during the second half of the S phase (late replication), G or R banding can be revealed after appropriate staining depending on the interval during which analogue incorporation occurs (Dutrillaux, 1975a). According to Holmquist (1989), each band corresponds to a group of replicons, the replication of which is initiated and terminated synchronously and at times which differ according to the type, G or R, of the bands. At present, it is not known whether this characteristic is related to the dispersed repeated sequences present in a band but it seems likely that it concerns all the DNA in a band.

In conclusion, it seems that in human chromosomes the banding methods yielding structural Q, G or R bands may be based on the differential distribution of SINE and LINE sequences along the chromosome. This research must however be extended to other species in order to generalize this model. It has been shown that sequences of this type are present in species other than man, such as B1 sequences of SINE type and MIF-1 sequences of LINE type in mouse (Singer, 1982b) and SINE type sequences in cattle (Lenstra et al., 1993), but they have not been precisely localised on chromosomes. It is more difficult to observe dynamic bands in non mammalian species and the difficulties encountered depend both on the method used and on the animal order. Concerning the animal order, our experience indicates that these difficulties increase as follows: placental mammals, marsupials, birds, reptiles, fish.

The following comments can be made on the influence of the banding techniques used:

- denaturation methods produce clear telomeric banding only in birds, it is less clear in reptiles and almost non existent in fish. Such methods also occasionally produce heterochromatin banding.

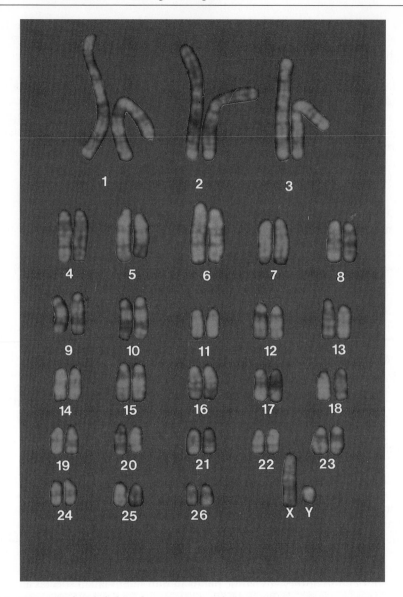

Figure II.1. RBP banded sheep karyotype classified according to the Texas 1995 nomenclature (photograph H. Hayes, reproduced by permission of CAB international).

- Q and G banding (using enzymatic digestion) gives varying, but more reproducible results.
- the techniques of choice are certainly those using BrdU incorporation. By their means, good results have been obtained even with reptiles and fish. The major problem is the difficulty of defining the optimal duration of

BrdU incorporation which varies through the cell cycle. Incorporation times varying from 7 to 18 hours give satisfactory results.

These observations suggest that "structural" banding elements are not directly associated with those of "dynamic" banding. They show also that replication heterogeneity, the existence of blocks of DNA sequences which replicate early or late appeared earlier during evolution than structural heterogeneity of chromatin. These DNA blocks are the basis of the bands in mammals and it has been observed in all species examined that the blocks which are replicated early are rich in genes and correspond to R and T bands.

II.1.3
Code of chromosome banding techniques

Chromosome banding thus permits the revelation of different structures or bands along the chromosomes. The numerous existing banding protocols are classified by a three letter code, published in the 1975 supplement to the proceedings of the 1971 conference on the standardisation of human chromosomes held in Paris (Paris Conference 1971, Supplement 1975).

In this code, the first letter indicates the type of banding, the second the banding technique and the third the stain used.

II.2
Techniques based on DNA structure

B. DUTRILLAUX and H. HAYES

A preliminary treatment during cell culture is not necessary in most techniques based on DNA structure.

Table II.2. Code for chromosome banding techniques

Code	Chromosome banding nomenclature
QFQ	Fluorescent Q bands stained by quinacrine
QFH	Fluorescent Q bands stained by Hoechst 33258
GTG	G bands by trypsin treatment stained with Giemsa
GAG	G bands by treatment with acetic saline stained with Giemsa
GBG	G bands by BrdU incorporation stained with Giemsa
CBG	C bands by barium hydroxide treatment stained with Giemsa
RFA	Fluorescent R-bands stained with acridine orange
RHG	R bands by heat denaturation stained with Giemsa
RBG	R bands by BrdU incorporation stained with Giemsa
RBA	R bands by BrdU incorporation stained with acridine orange
RPB	R bands by BrdU incorporation stained with propidium iodide

II.2.1
QFQ bands using quinacrine mustard

Principle

Q banding described by Caspersson and colleagues in 1968 was the first chromosome banding method to be developed. The Q bands (Figure II.2) are produced by staining with quinacrine mustard, a fluorochrome which binds to DNA either by intercalation between the planes of successive base pairs or by ionic bonds. Binding occurs preferentially with AT base pairs and the fluorescence of quinacrine mustard bound to AT base pairs is increased compared to that of the same compound bound to GC base pairs (Weisblum and de Haseth, 1972). These characteristics may constitute the basis of the Q

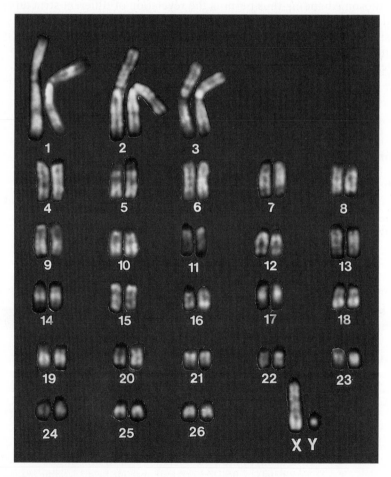

Figure II.2. QFQ banded sheep karyotype classified according to the Texas 1995 nomenclature (photograph H. Hayes).

banding pattern since the AT base pair content of DNA varies along the chromosomes (see paragraph II.1.2.1). However, this hypothesis does not explain why quinacrine mustard fails to stain certain AT base pair rich regions of heterochromatin.

Protocol
The slides are incubated for 10 minutes at room temperature in 1× quinacrine mustard solution (appendix S9), rinsed with Sorensen's buffer pH 6.8 (appendix B3) and mounted in the same buffer. The slides are examined under a fluorescence microscope using the filter block appropriate for quinacrine mustard.

Remarks
Chromosomes show a series of dark and brightly fluorescent regions of different sizes. The intensity of fluorescence gradually diminishes during exposure to ultraviolet light and for this reason, the bands should be photographed rapidly. This banding method possesses the advantage of not altering the structure of chromosomes and therefore can be used for sequential treatment of slides (see paragraph II.7). For example, after identification of chromosomes by Q banding, the slides can be destained and a second banding method can be applied, or they can be used for *in situ* hybridisation.

II.2.2
GTG banding by trypsin

Principle
GTG bands (Figure II.3) are obtained by the action of trypsin on the chromosomes. The mechanism by which trypsin reveals the GTG banding pattern, briefly discussed in paragraph II.1.2.1 is at present poorly understood. The technique described here is that of Seabright (1971).

Protocol
The slides are rinsed in distilled water, incubated for 10 to 20 seconds at room temperature in a freshly prepared 0.25% trypsin solution (appendix S5) and then washed twice in succession in baths of PBS⁻ (appendix B1) in order to block the action of trypsin. They are then stained with Giemsa and examined under the microscope in normal light with a green filter.

Remarks
GTG and QFQ band patterns are almost identical and the former has the advantage of being stable. Nevertheless, the GTG banding pattern is not easily reproducible and depends on the quality of preparations and the age of slides. After spreading, slides should generally be stored for one week, before the GTG banding procedure and the duration of trypsin treatment must always be adjusted. If this treatment is too short, the chromosomes are entirely stained and if it is too long, the chromosomes are weakly stained and

Figure II.3. GTG banded gazelle karyotype (*Gazella subgutturosa marica*) (photograph M. Vassart). In gazelle, the X chromosome is fused with the homoeologous cattle chromosome 5 and thus the nomenclature of this male is X, Y2 (homologue of homoeologous cattle chromosome 5) and Y1 (chromosome Y). Moreover, this karyotype is that of a heterozygote carrier of a Robertsonian translocation numbered 14 and composed of chromosomes homoeologous to cattle chromosomes 1 and 3 (Vassart et al., 1995).

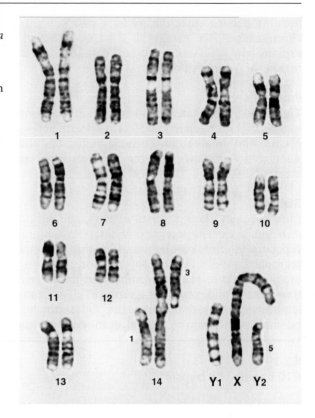

their morphology is altered. This technique has been extensively used, especially for producing the first standard karyotypes of many species. The centromeric regions of bovine chromosomes are usually very weakly stained which can make their classification difficult. In contrast, in man these regions are strongly stained and this differentiates GTG and QFQ banding.

II.2.3
GAG banding by "denaturation"

Principle
Many G banding methods involving saline treatments have been described, but most are now obsolete and have been replaced by simpler enzymatic digestion methods.

Two main methods were described in 1971:

- Acid-Saline-Giemsa by Sumner et al. (1971)
- Alkaline-Saline-Giemsa by Schnedl (1971)

Protocol Acid-Saline-Giemsa

The slides are incubated in 2× SSC buffer pH 7 (appendix B4) for one hour at 60 °C. They are then rinsed with distilled water and stained for 90 minutes in a 2% Giemsa solution (see paragraph I.5.1). The preparations are observed under the microscope in normal light, with a green filter.

Protocol Alkaline-Saline-Giemsa

The slides are incubated in 0.07 M NaOH (appendix SS12) for 90 seconds at room temperature, then dehydrated in three successive baths of 70%, 90% and 100% ethanol. They are then incubated for 24 hours in Sorensen's buffer pH 6.8 (appendix B3), at 59 °C and finally stained for 10 minutes with 10% Giemsa at pH 6.8 (see paragraph I.5.1).

Remarks

The large number of "modified" or "improved" versions of these techniques which have been proposed by different laboratories since 1971 should all be considered simply as local variants valid for one laboratory but not necessarily for others. It is always preferable to use the original techniques and to adapt them by modifying the duration of treatment in 2× SSC buffer or in sodium hydroxide solution, or the conditions of Giemsa staining. If the preparation is too strongly stained, the duration of treatment with sodium hydroxide should be increased or that of staining reduced. On the contrary, if C-like banding appears, the sodium hydroxide treatment may be reduced or suppressed or the time of staining increased.

The treatments used to obtain chromosome preparations play an important role in the quality of the G banding (GTG or GAG). Thus, the hypotonic treatment using diluted serum described earlier, is the best method for obtaining R bands, but is not satisfactory for G band revelation for which it is preferable to use a hypotonic treatment with KCl and a fixation with "methanol – acetic acid" instead of "ethanol – acetic acid".

Many failures are probably caused by not respecting these conditions. In the end, laboratories choose more or less deliberately the banding technique best adapted to the cytological methods they employ rather then the best available technique.

II.2.4
RHG banding by thermal denaturation

Principle

RHG banding is based on the thermal denaturation of chromosomes but in this case also, the biochemical mechanism of band formation is not well understood. After denaturation of the chromosomal DNA under precise conditions of temperature and pH only regions rich in GC base pairs remain undenatured and are stained by Giemsa (Comings, 1978). A modification of the technique first described by Dutrillaux and Lejeune (1971) is described here.

Protocol

The slides are incubated at 87 °C for 30 to 90 minutes according to their age in 1× bicarbonate free Earle's solution at pH 5.3 (appendix B9) and then at 87 °C for 30 minutes in 1× bicarbonate free Earle's solution at pH 6.5 (appendix B9) after which they are rinsed in running water and stained with Giemsa (see paragraph I.5.1). Observation is carried out under the microscope in phase contrast with an orange filter.

Remarks

RHG banded chromosomes (Figure II.4a and 4b) show a series of more or less strongly stained bright and dull fluorescent and more or less wide bands the sequence of which is the inverse of that of Q and G bands. In bovidae chromosomes, eg sheep chromosomes (Figure II.4b), the centromeric regions are strongly stained, particularly in the case of acrocentric chromosomes and sometimes the banding is similar to CBG banding. This indicates that the constitutive heterochromatin of the centromeric regions of these species is rich in GC base pairs. Intense staining of the centromeric regions by the RHG banding technique is typical for bovidae chromosomes. It is observed neither in human chromosomes, nor in those of most other mammals. This technique also stains the telomeric regions of almost all chromosomes and thus unlike the QFQ or GTG banding methods it has the advantage that it reveals the total length of each chromosome.

The appearance of RHG bands corresponds to a stage of chromosome denaturation. The rapidity with which this stage is reached depends on the age of the preparation. The technique described above is suitable for "fresh" 1 to 3 day old slides. For older slides, the time of treatment should progressively decrease to about 20 minutes for 15 to 20 day old slides and to less than 1 minute for slides prepared several months before treatment. The range of treatment times within which best quality staining is obtained also decreases in parallel with the age of preparation passing from several minutes for fresh slides and to few seconds for old slides. It is therefore better to work with fresh slides, or avoid the ageing of preparations by storing slides at or below −20 °C. Slides used for *in situ* hybridisation should be stored in the same way since this technique also involves DNA denaturation.

If a slide is too strongly stained after the treatment or if the chromosomes appear feathery or if a weak G type banding is observed, this indicates insufficient treatment. The same preparation can be submitted to a second treatment after removing immersion oil by rinsing successively in several (normally 6) toluene baths. It is not necessary to destain the slide but this can be done by washing in a bath of 70% ethanol.

If a preparation is too "denatured", the chromosomes will present the following aspects in the indicated order: very weak R bands, T bands, C bands and finally complete destaining.

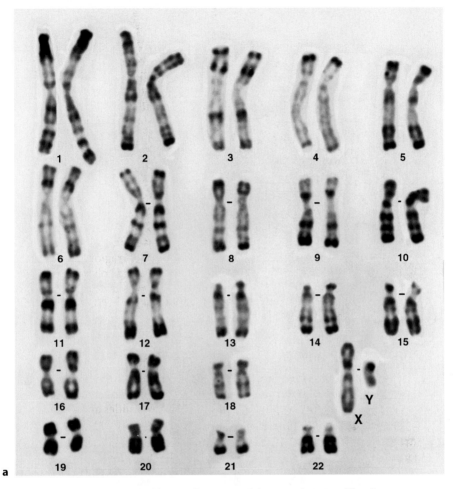

Figure II.4. (a) RHG banded human karyotype (photograph B. Dutrillaux).
(b) RHG banded sheep karyotype classified according to the Texas 1995 nomenclature
(photograph H. Hayes). Note the strong staining of the centromeric regions in the sheep
karyotype.

II.2.5
T bands (terminal)

B. Dutrillaux and P. Popescu

Protocol
The technique of T banding is very similar to that of R banding, except for
the pH of the 1× Earle's solution (appendix B9) which should be more acid,
between 5.1 and 5.3. This solution is used alone as a denaturing bath at 87 °C
for a duration in inverse ratio to the age of the slides: 90 to 75 minutes for

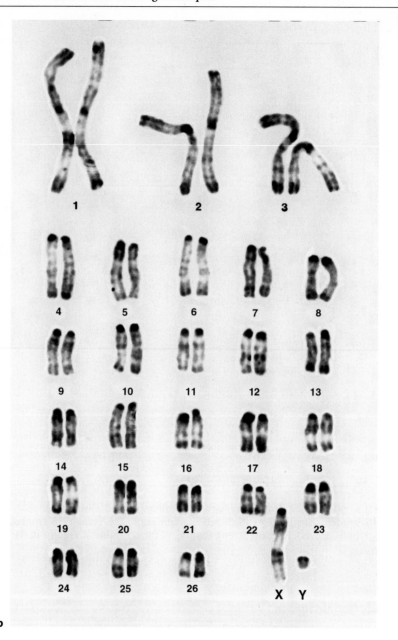

b

Figure II.4. *Continued.*

one day slides and about 20 minutes for 15 day old slides. Although Giemsa staining can be used, the results are clearer after staining with acridine orange (Figure II.5).

Remarks
T bands correspond to a subclass of R bands containing those most resistant to thermal denaturation. Most but not all T bands are situated in the terminal regions of chromosomes (Dutrillaux, 1973; Ambros and Sumner, 1987; Mezzelani et al., 1996) which are the chromosomal segments richest in GC base pairs and in genes (Saccone et al., 1992 and 1993; Gardiner, 1996).

In pig, T banding occurs in the centromeric regions of a subgroup of chromosomes, the meta- and submetacentric chromosomes. Pig chromosomes possess two different types of satellite DNA in their centromeric region: a repeated sequence of about 100 base pairs is present in the pericentromeric regions of the 12 metacentric and submetacentric chromosomes and another repeated sequence of 14 base pairs in the pericentromeric regions of the acrocentric chromosomes (Jantch et al., 1990). In contrast to C banding, which stains uniformly the centromeric regions of all pig chromosomes, T banding as well as staining telomeric regions, stains the centromeric regions only of the meta- and submetacentric chromosomes. This particularity of T banding in the pig is very useful for the identification of reciprocal translocations involving a meta- or submetacentric chromosome and an acrocentric chromosome. It permits detection of the transfer of a part of the satellite DNA present in the centromeric region of a meta- or submetacentric chromosome to an acrocentric chromosome (Figure II.6, Gustavsson and Settergren, 1984; Popescu et al., 1984).

II.2.6
Bands rich in 5-methylcytosine

C. BOURGEOIS

II.2.6.1
Introduction

Nucleotide methylation is the only epigenetic modification of DNA at present known in eucaryotes. Of the four DNA bases, essentially only cytosines present in CpG dinucleotides are methylated in position 5 (5-methylcytosine or 5-mC). The percentage of 5-mC in DNA is very variable according to species and tissue. It is generally higher in plants than in animals and in some cases it can be close to zero, for example Drosophila. The importance of methylation for the activity of certain genes and its implication in differentiation, in embryogenesis and genomic fingerprinting are well known although its exact role has not been elucidated.

Many procedures have been used to study DNA methylation: chemical analysis, molecular biology and immunofluorescence of chromosomes. Chemical analysis gives a global vision of methylation but does not permit

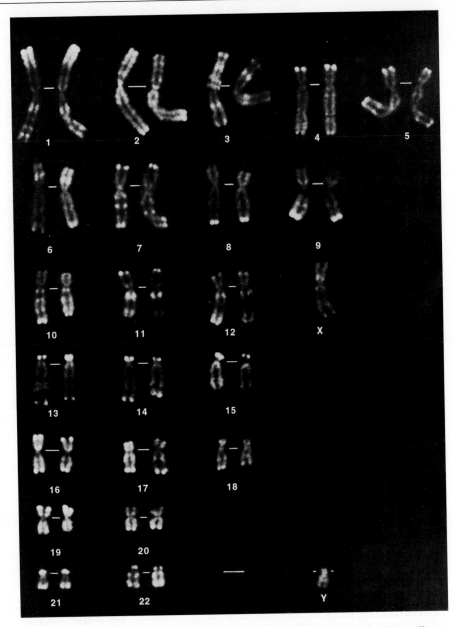

Figure II.5. T banded human karyotype stained with acridine orange (from Dutrillaux, 1975b, reproduced by permission of Expansion Scientifique Française).

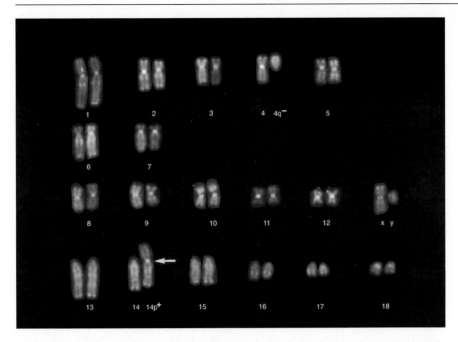

Figure II.6. T banded pig chromosomes stained with acridine orange. Note the translocation between chromosomes 5 and 14 and the transfer of a part of the centromeric DNA of chromosome 5 to chromosome 14 (from Popescu et al., 1984, reproduced by permission of the American Genetic Association).

identification of the methylated regions of the genome. Molecular biological studies use restriction enzymes sensitive to nucleotide methylation: after digestion of DNA by such an enzyme the products are analysed by electrophoresis on an agarose gel and transferred to a nylon membrane according to Southern (1975). After hybridisation with appropriate probes and autoradiography, it is possible to compare the migration profiles of the DNA of different species, of different tissues, or taken at different stages of development. This type of analysis gives information which is precise but limited to 4–8% of the genome since it involves only the DNA sites recognised by each restriction enzyme. Recently, an immunofluorescent method applicable to metaphase chromosomes has been described (Barbin et al., 1994). This method described below, permits a global vision of the degree of methylation and of the modifications which this induces, chromosome by chromosome and band by band.

II.2.6.2
Immunofluorescent revelation of 5-mC rich bands

Principle
Antibodies specifically raised against 5-mC are used for detecting the methylated sites along the chromosome. After denaturation of the chromosome

preparations fixed on slides, these antibodies bind to the methylated sites in single stranded DNA. They are detected by a second antibody raised against the first antibody and conjugated to a fluorochrome.

Protocol

The procedure involves two steps:

- denaturation is necessary to increase the accessibility of methylated sites to anti-5mC antibodies in order to maximise the number of sites detected and the reproducibility of their detection. Two protocols are described below.
- immunofluorescence, reaction with antibodies and detection.

If frozen slides (stored at −20 °C) are used, they are kept at room temperature for 72 hours before use. Before denaturation of preparations, slides are rehydrated in phosphate buffer pH 7.3 (appendix B10).

■ **Denaturation step.** Denaturation can be accomplished either by acid treatment, or by ultraviolet radiation.

Protocol for denaturation with hydrochloric acid

Chromosome preparations, obtained either from lymphocytes or fibroblasts are first submitted to a proteolytic treatment in a solution of pepsin (20 µg/ml 0.1 N HCl) at 37 °C for 10 to 15 minutes according to the age of the slides and to the species analysed. The slides are then rinsed in phosphate buffer pH 7.3 to block action of the pepsin, and airdried at room temperature for 30 minutes after which they are incubated in a bath of 2 N HCl at 37 °C for 10 to 20 minutes. The number of sites observed and the morphology of chromosomes depend critically on the length of this treatment. After acid treatment, the slides are immediately rinsed twice in succession for 5 minutes at 4 °C first in a bath of borate buffer pH 8.4 (appendix B8) and then in distilled water. They are then dehydrated in three successive baths (5 minutes each at 4 °C) of 50, 75 and 100% of ethanol.

Protocol of denaturation using ultraviolet radiation

This technique is described by Schreck and colleagues (1974). The slides are placed in a dish, covered with phosphate buffer pH 7.3 (appendix B10) and exposed to an ultraviolet source using a 30 W germicidal lamp or two 15 W lamps (Philips ultraviolet 15 W) at a distance of 30–32 cm for 8 to 14 hours. They are then rinsed with phosphate buffer pH 7.3 before treatment with antibodies.

■ **Immunofluorescence revelation.** After denaturation, slides are stored at room temperature for 24 hours, rehydrated in phosphate buffer pH 7.3 (appendix B10) for 5 minutes and incubated in PBT (appendix H3) for 10 minutes. 100 µl of anti-5mC antibody solution diluted in PBT (the optimal concentration of antibody is determined for each group of slides) are deposited on each slide and covered with a coverglass and the slides are incubated at 37 °C for 45 minutes in a humid chamber in the dark. After rinsing in phosphate buffer pH 7.3 for 5 minutes and then PBT for 10 minutes, 100 µl of a solution of the antibody raised against the first antibody and conjugated to rhodalgreen are placed on each slide (rhodalgreen is preferred as its fluorescence decreases more slowly and is more intense than that of fluorescein). The slides are incubated again as described above. Some minutes before the end of the incubation with the second antibody, the preparations can be counterstained by adding 100 µl of 1× propidium iodide solution at a final concentration of 1 ng/µl (appendix S12) to each slide and continuing the incubation for 6 minutes at room temperature in the dark. Since the first step of this technique causes complete denaturation of the DNA double helix, it may be necessary to increase the concentration of propidium iodide to obtain optimal staining.

Chromosome preparations (Figure II.7) are observed in presence of PPD8 (appendix S11) using a fluorescence microscope equipped with a double transmission band filter system (fluorescein/rhodamin excitation) to permit the simultaneous observation of the green fluorescence emitted by rhodalgreen and the red fluorescence emitted by propidium iodide.

Remarks

Human chromosomes display bands of different fluorescence intensity (bright and dull) along their length and this permits the construction of a map of 5-mC rich sites. The variation of the intensity of fluorescence of these bands is related to the different 5-mC contents of chromosomal regions.

Four classes of sites varying in their richness in 5-mC can be distinguished:

- type I, the richest, corresponds to regions containing the classical DNA satellites
- type II is situated in T bands and in one or two neighbouring sites such as, for example band q43 of human chromosome 1 which contains the genes for the proteins of the 5S ribosomal subunit
- type III is situated in R bands (G bands show a very weak fluorescence intensity)
- type IV occurs in the short arms of the acrocentric chromosomes. The fluorescence intensity of these sites varies according to the individual and it may be connected to the presence of the ribosomal genes in these regions.

Denaturation using ultraviolet radiation permits a better preservation of chromosomal morphology, but the detection of the different types of site varies according to the duration of exposure to ultraviolet light. For example, in man, the heterochromatic regions appear first, their fluorescence intensity increases initially, then decreases and finally disappears whereas the fluorescence intensity of telomeric bands continues to increase. Such behaviour probably also accounts for the heterogeneity of banding observed in different metaphase spreads on the same slide.

This method can be used, without modifications, to analyse the dynamics of DNA methylation during meiosis in man and animals.

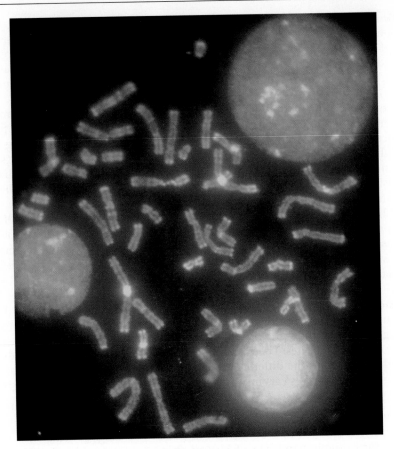

Figure II.7. Human chromosome spread marked by the technique of bands rich in 5-methyl cytosine revealed by staining with rhodal green (green fluorescence) and counterstaining with propidium iodide (photograph C. Bourgeois).

II.3
Banding techniques based on DNA replication

B. DUTRILLAUX and H. HAYES

II.3.1
R or G bands by incorporation of BrdU

Principle
Chromosome banding methods based on DNA replication require preparations of chromosomes in which the DNA has been modified *in vitro* during its replication, by incorporation of 5-bromo-2-deoxyuridine, BrdU, a structural analogue of thymidine.

In such preparations, differential staining of chromosomal regions, the DNA of which does or does not contain BrdU can be achieved either by:

- immunodetection using anti-BrdU antibodies or by
- staining procedures which exploit the photosentivity of chromosome segments containing incorporated BrdU.

As explained earlier, two euchromatin fractions (Figure II.8) can be distinguished:

- in one, located in R bands, the DNA replicates during the first half of the S phase (early replication)
- in the second, located in Q and G bands, the DNA replicates during the second half of the S phase (late replication).

This characteristic can be used to obtain one or other of these band patterns along the chromosomes. The presence of BrdU during the final hours of a culture (Figure II.8) leads to its incorporation in late replicating euchromatin. After appropriate treatment it is then possible to stain only regions which do not contain incorporated BrdU, and to reveal an R banding pattern. Similarly, if BrdU incorporation occurs before the final phase of a culture ie in early replicating euchromatin a G band pattern can be observed.

The duration of BrdU incorporation for different types of cultured cells must be determined by preliminary trials. However, if unsynchronized cultures are used, this method produces a mixture of cells with chromosomes

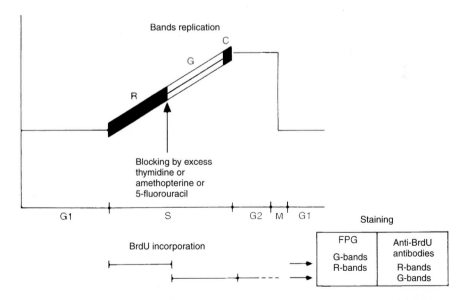

Figure II.8. BrdU incorporation times during the cell cycle which produce "dynamic" R or G bands. G_1, S (DNA synthesis), G_2 and M (mitosis) are the different phases of the cell cycle.

containing various degrees of BrdU incorporation. It is therefore preferable to use synchronised cultures (see paragraph I.2.1) in which most cells are at the same stage of the cell cycle. BrdU incorporation is almost equivalent for all cells. Band patterns then observed are reproducible and characteristic for the stage of the cell cycle at which BrdU is added and for the duration of its incorporation. Before discussing protocols for BrdU incorporation, the major chromosome staining techniques are presented with the exception of the use of acridine orange already described in paragraph I.5.3.

II.3.1.1
Immunofluorescent detection of BrdU incorporation in chromosomes

Principle
The production of monoclonal anti-BrdU antibodies which are now commercially available led to the development of immunodetection techniques to localize BrdU incorporated *in vitro* in chromosomes (Vogel et al., 1986). To permit the detection of antigenic sites, denaturation of DNA is required, although the mechanism of fixation of anti-BrdU antibodies on DNA is not yet elucidated, it has been shown that denaturation of DNA is necessary to make the antigenic sites accessible. In fact, the antibodies cannot bind to double stranded DNA and the condensed chromatin of metaphase chromosomes requires more severe denaturation conditions than the chromatin of interphase chromosomes. Different treatments have been described (Drouin et al., 1989) but only two, the choice between which depends on the type of information desired, are described here, namely:

– the use of hydrochloric acid for direct revelation of anti-BrdU antibodies
– the use of formamide for revelation following *in situ* hybridisation.

The immunodetection can be performed either with a single fluorescein conjugated antibody or with a system of two antibodies: the first raised against BrdU and the second raised against the first antibody and conjugated with a fluorochrome. The selection of the technique to be used will depend on the extent of incorporation of BrdU in chromosomes: if the BrdU concentration added to cell cultures is >5 µg/ml and the incorporation time >30 minutes, the method using one antibody is sufficiently sensitive. On the contrary, for lower concentrations of BrdU and/or shorter incorporation times, the method using two antibodies which yields higher signal intensities is preferable.

Protocol
The slides are rehydrated in phosphate buffer (appendix B10), then treated gently with pepsin by incubation for 5 to 10 minutes at 37 °C in a solution of pepsin, 50 µg/ml in 0.1 N HCl, after which the action of pepsin is blocked by several brief rinses in phosphate buffer and the slides are airdried for 30 minutes. The chromosomal DNA is then denatured by treatment either with hydrochloric acid or with formamide, as noted above.

■ **Denaturation using hydrochloric acid.** The slides are incubated successively in 2 N HCl for 2–4 minutes at room temperature, in borate buffer pH 8.4 (appendix B8) for 5 minutes at 4 °C and in distilled water for 5 minutes at 4 °C.

■ **Denaturation using formamide.** The slides are incubated one by one for 3 minutes at 70 °C in 2× SSC buffer pH 7 (appendix B4) containing 70% deionised formamide (appendix M7), then immediately rinsed twice in succession for 2 minutes each at 4 °C in 2× SSC buffer pH 7.

After denaturation by either of the methods described here, the slides are dehydrated by rinsing successively for 2 minutes at 4 °C in 50%, 75% and 100% ethanol and airdried for at least one hour and at most three days.

■ **Immunodetection.** After rehydration in phosphate buffer (appendix B10) for 5 minutes at room temperature, slides are incubated in PBT solution (appendix H3) for 10 minutes. 100 µl of a PBT solution containing anti-BrdU antibody diluted according to the supplier's instructions are then deposited on each slide, coverslips are added and the slides are incubated at 37 °C for 45 minutes in a humid chamber. If the anti-BrdU antibody is directly conjugated to a fluorochrome, slides are rinsed twice for 5 minutes in phosphate buffer and mounted with the same solution containing an antifading agent such as PPD7 phenylenediamine at pH 7 (appendix S11). If the anti-BrdU antibody is not directly conjugated to a fluorochrome, slides are rinsed at room temperature in two successive baths of phosphate buffer (5 minutes each), then in PBT. 100 µl of a PBT solution of the second fluorochrome conjugated antibody, at the concentration recommended by the supplier, is then placed on slides and they are incubated, rinsed and mounted as already described.

Preparations can be counterstained with propidium iodide (paragraph II.3.1.3) or DAPI (paragraph II.3.1.4) before mounting.

Remarks
Chromosomes display patterns of fluorescent and non fluorescent bands similar to those obtained by other staining techniques (FPG, acridine orange, propidium iodide, DAPI). This technique permits the detection of BrdU incorporated into the chromosomes of cells subjected to a discontinuous BrdU treatment and, therefore, the analysis of the chronology of DNA replication along the chromosome.

II.3.1.2
FPG staining technique (fluorochrome-photolysis-Giemsa)

Principle
Chromosome preparations are stained with the photosensitive coloring agent Hoechst 33258 which binds to DNA and then irradiated with "black" light (Mazdafluor OETFWN 20 lamp) at a wave length corresponding to that of the excitation range of Hoechst 33258. The combination of Hoechst 33258 and "black" light causes photolysis of BrdU in DNA and a series of radical

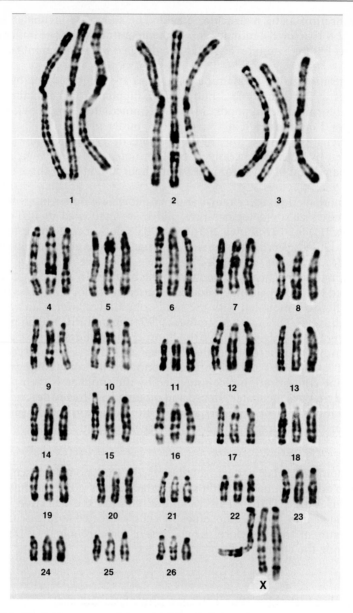

Figure II.9. Haploid montage of three RBG banded sheep karyotypes classified according to the Texas 1995 nomenclature (photograph H. Hayes).

reactions, the effects of which reduce the affinity of chromosome regions containing BrdU rich DNA for the Giemsa stain. Consequently, only the DNA regions which did not incorporate BrdU will be stained in chromosomes (Figure II.9). Studies performed with anti-BrdU antibodies (Drouin et al.,

1989) showed that this differential staining is due to the degradation and selective extraction of DNA regions containing incorporated BrdU. Several variants of the original protocol published by Perry and Wolf (1974) exist; that described below is used in Dutrillaux's laboratory.

Protocol
The slides are incubated for 20 minutes at room temperature in a freshly prepared 1× Hoechst 33258 solution (appendix B6) and then rinsed in water. They are then placed in dishes, covered by 2× SSC buffer pH 7 (appendix B4) and exposed to black light at a distance of 10 cm for 90 minutes. After rinsing in water, they are incubated at 87 °C for 10 minutes in 1× bicarbonate free Earle's solution pH 6.5 (appendix B9), rinsed and stained with Giemsa (see paragraph I.5.1). The observation is carried out under the phase contrast microscope with an orange filter.

Remarks
The chromosomes are more or less strongly stained according to the extent of BrdU incorporation. A predominance of dark bands indicates that the duration of BrdU incorporation was too short and a predominance of bright bands that it was too long.

II.3.1.3
Propidium iodide staining technique

Principle
This technique was developed by Lemieux and colleagues (1992) to permit simultaneous observation by fluorescence microscopy of RPB or GBP bands (paragraph II.3.1) and the hybridisation signals in the chromosomes treated by *in situ* hybridisation (Chapter III). In this technique, the slides are stained with propidium iodide after hybridisation and then observed in the presence of PPD at a very alkaline pH (the antifading agent PPD p-phenylenedyamine is normally used at neutral pH). Under exposure to blue excitation light, chromosome regions which do not contain incorporated BrdU emit a red fluorescence, while the regions which do contain incorporated BrdU fade very quickly and appear dull (Figure II.10). The mechanism of this differential staining is not yet understood. However, the modification of chromatin structure due to the different treatments applied to chromosomes during *in situ* hybridisation and in particular the alkaline conditions during staining, and observation under the fluorescence microscope are important factors in the process.

Protocol
100 μl of a propidium iodide solution at 1 μg/ml (appendix S12) is placed on slides and they are incubated in the dark for 6 minutes at room temperature. They are then rinsed in phosphate buffer pH 7.3 (appendix B10) and mounted with a coverslip in 10 μl of PPD 11 (p-phenylenediamine at pH 11;

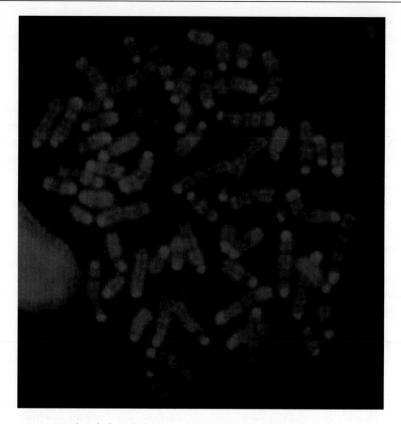

Figure II.10. RBP banded cattle karyotype (photograph H. Hayes).

appendix S10) and examined under the fluorescence microscope using a filter system appropriate either for rhodamine or for fluoroscein isothiocyanate (FITC) permitting in the case of *in situ* hybridization, simultaneous observation of the green fluorescence (FITC) of the hybridization signals and of the red fluorescence (propidium iodide) of the banding pattern.

Remarks

As in the FPG staining technique, chromosomes are stained more or less strongly according to the extent of the incorporation of BrdU (Figure II.10). In addition, the staining intensity varies from one chromosome preparation to another and according to the species from which the chromosomes are derived. Sometimes stain concentration must be adjusted slide by slide. This technique is simple and very effective for the production of R or G bands after *in situ* hybridisation (Figure II.11).

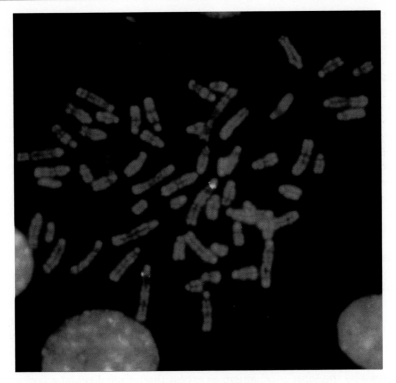

Figure II.11. RBP banded cattle chromosome spread after *in situ* hybridisation with a GRIK1 gene biotinylated probe from a bovine YAC (photograph H. Hayes). Hybridisation signals are visible on band q14 of each cattle chromosome 1.

II.3.1.4
DAPI staining technique

Principle
DAPI (4'-6-diamidino-2-phenylindole) is a fluorochrome which has an affinity for AT base pairs and it therefore produces a banding pattern similar to QFQ bands. Rarely used alone it is usually associated with a counterstain such as distamycin A (see paragraph II.5.4). This association intensifies the staining of certain regions.

Protocol
The slides are incubated for 15 to 30 minutes in the dark in MacIlvaine's buffer pH 7 (appendix B6) containing DAPI at a concentration of 0.4 µg/ml (appendix S2). They are then rinsed and mounted in MacIlvaine's buffer pH 7. They are observed under the fluorescence microscope using a filter system appropriate for DAPI.

II.3.1.5
Preparation of chromosomes labeled with BrdU during the second half of the S phase to produce R bands

Protocol
The protocol described here derives from those published by Dutrillaux and colleagues (1973) and by Viegas-Péquinot and Dutrillaux (1978).

To a cell culture synchronised by thymidine or amethopterin, after release of the block on the day cells are to be harvested, 50 µl of 1× BrdU solution (appendix V5) is added (final concentration 10 µg/ml). One hour later, 100 µl of 1× FdU solution (5-fluoro-2′-deoxyuridine; appendix V6) is added (final concentration 0.5 µg/ml) to the same cultures. Cells are then incubated, harvested and treated for the preparation of spreads according to paragraph I.4. RBG and RPB bands are produced by the techniques which have already been described.

Remarks
In this case, since thymidine and amethopterin block the cell cycle in the middle of the S phase, BrdU is incorporated into DNA from the moment the cell cycle resumes ie during the second half of the S phase (Figure II.8). Hence, only DNA which was already synthesised before incorporation of BrdU could occur, will be stained after treatment and an R band pattern is therefore obtained. The final concentration of BrdU in the culture medium is an important parameter; BrdU concentrations above 300 µg/ml block DNA synthesis by the same mechanism as excess thymidine. Even at a concentration of 10 µg/ml as used in the protocol above, BrdU slows down the cell cycle and the interval of time between release of the mitotic block and the onset of mitosis is increased by about one hour compared to that in the absence of BrdU. If BrdU concentrations below 10 µg/ml are used, incorporation is insufficient and the chromosomes are weakly stained. BrdU incorporation can be increased by blocking endogeneous thymidine synthesis, by addition of FdU which is an inhibitor of thymidylate synthetase.

The banding patterns obtained by the FPG technique (RBG bands) or by the propidium iodide technique (RBP bands) are generally of very high quality. The bands are more numerous and sharper than those produced by techniques based on DNA structure. Banding patterns obtained in this way have proved very useful for the detailed analysis and improved identification of mammalian and avian chromosomes. The two X chromosomes in spreads of female mammalian cells can be differentiated: the active X chromosome is stained like an autosome since its DNA is replicated throughout the whole of the S phase whereas the inactive X chromosome is completely unstained since its DNA is replicated at the end of the S phase.

Incorporation of BrdU into DNA varies significantly from one type of cell to another in relation to the activity of thymidine kinase. In man, BrdU incorporation is readily obtained in lymphocytes, but less easily in fibroblasts. In some cancer cells, degradation or very rapid uptake of BrdU can occur leading to paradoxical banding patterns which may be incorrectly interpreted. For example, a continuous treatment with BrdU can produce the same banding pattern as a short discontinuous treatment (Massad et al., 1991).

II.3.1.6
Preparation of chromosomes labeled with BrdU during the first half of the S phase to produce G bands

Protocol
To obtain GBG or GBP banding, cells are blocked by amethopterin according to the protocol described in paragraph I.2.2 but in the presence of BrdU which is added to the culture medium (final concentration 10 µg/ml; appendix V5) at the same time as amethopterin. After release of the block, culture is continued in the absence of BrdU and the rest of the procedure is as described in paragraph I.4. GBG or GBP bands are then stained by the techniques previously described.

Remarks
Under the above conditions, BrdU is incorporated until DNA synthesis is blocked by amethopterin (Figure II.8) and thus only the DNA synthesises after the cell cycle resumes ie that located in G bands will be stained after treatment. The quality of the GBG and RBG banding patterns obtained is similar.

It is also possible to treat cultures with a high concentration of BrdU (200 µg/ml); incorporation takes place during the first half of the S phase followed by a block in the middle of the S phase. After rinsing, the cells are resuspended in standard culture medium for 6 to 7 hours.

II.3.2
Sister chromatid exchanges (SCE)

Principle
BrdU incorporation during two consecutive cell cycles leads to different BrdU distributions on sister chromatids, because of the semi-conservative replication of DNA. After one cycle of replication in the presence of BrdU, one strand of each DNA molecule contains incorporated BrdU and the other does not. After a second cycle of replication in the presence of BrdU only one strand in four will not contain incorporated BrdU and each chromosome will contain one chromatid with BrdU in both strands and one with BrdU in only one strand. The two chromatids therefore stain unequally, the former more weakly than the latter and so it can be differentiated. Furthermore, because of this difference in staining, transfer of segments from one chromatid to the other (sister chromatid exchanges) become visible.

Protocol
Cells are cultured as described in Chapter I. 40 to 48 hours before the end of the culture, a solution of BrdU (appendix V5) is added at a final concentra-

tion of 10 µg/ml. Cells are harvested and chromosome preparations are made according to the usual protocols. Chromosome spreads can be stained either by acridine orange (paragraph I.5.3) or by the FPG techniques (Figure II.12; paragraph I.3.1.2).

Remarks

Sister chromatid exchanges are detected in many species including man. Their number is proportional to the concentration of BrdU used. With human cells and a BrdU concentration of 10 µg/ml, an average of 5 to 8 exchanges is observed per cell. Sister chromatid exchanges during the S phase appear to be related to DNA repair but their precise role is not understood. This technique is used mainly to test the effect of mutagenic substances (Perry and Evans, 1975) by comparing the frequency of exchanges in their presence and absence.

II.3.3
Asymmetrical incorporation of BrdU

In certain chromosomal regions (constitutive heterochromatin), asymmetrical uptake of BrdU into the two chromatids is observed after incorporation during a single cell cycle. These regions are always located in constitutive heterochromatin. In man, this asymmetry is observed in the Y chromosome, in

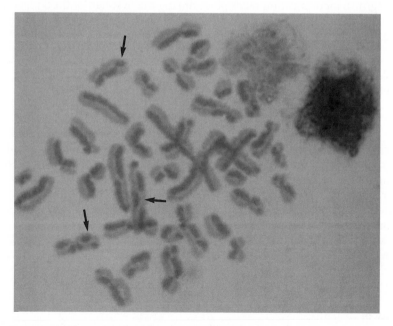

Figure II.12. Pig chromosome spread stained by the FPG technique and showing sister chromatid exchanges (SCE) (photograph P. Popescu).

the short arms of the acrocentric chromosomes and in the secondary constrictions of chromosomes 1, 9 and 16. In mouse, all the centromeric regions possess this characteristic. To explain this phenomenon, Lin and colleagues (1974) proposed that the asymmetry reflects that of the distribution of A and T nucleotides in the two complementary strands of repetitive DNA. Thus, during replication the newly formed strand complementary to the A rich strand will be rich in BrdU whereas the newly formed strand complementary to the T rich strand will contain mainly A and little BrdU. For this reason, one daughter chromatid contains normal DNA and the other DNA in which T is extensively replaced by BrdU.

II.4
Techniques of chromosome differentiation based on DNA base composition

B. DUTRILLAUX

II.4.1
Treatment by 5-azacytidine or 5-azadeoxycytidine

Principle
5-azacytidine (5-ACR) is an analogue of cytidine. As soon as it enters cells, it is quickly reduced to deoxy-5-azacytidine (5-dACR) before incorporation into DNA during replication. 5-ACR and 5-dACR can therefore be used interchangeably. After incorporation, 5-dACR inhibits the activity of the methylase which normally methylates about 5% of newly incorporated deoxycytidine residues. This has a profound effect on chromosome compaction notably in certain specific segments: G bands and especially constitutive heterochromatin containing classical satellite DNA sequences.

Protocol
5-ACR or 5-dACR is used at a final concentration of 10^{-5} M (Viegas-Péquinot and Dutrillaux, 1976). A 10× solution of 5-ACR or 5-dACR (10^{-4} M, appendix V7 and V8) is added to a cell culture, at a final concentration of 10^{-5} M, for the last 7 hours of culture. Cells treated for this length of time show extended chromosome segments, corresponding to G bands and to heterochromatin and R type banding is obtained. The alteration of chromosome compaction is so great that the banding is observed after simple Giemsa staining (see paragraph I.5.1), without any other treatment.

Remarks
Though the result of this method is spectacular, it is not much used in routine cytogenetics, compared with methods using BrdU which are better known. Like these methods it also permits differentiation of X chromosomes in which DNA is replicated early and late in the cell cycle. It also gives some indications concerning the composition of constitutive heterochromatin

since two types of heterochromatin are distinguished using this method: one, sensitive to 5-ACR incorporation because rich in methylatable CpG sites, the other insensitive (compaction not altered by treating cells with 5-ACR) because poor in such sites.

II.5
Heterochromatin staining

B. DUTRILLAUX and P. POPESCU

II.5.1
CBG bands

Principle
The C banding technique selectively stains constitutive heterochromatin at all stages of the cell cycle. In mitotic chromosomes, C bands are mostly localised in the centromeric regions, but they can be observed elsewhere depending on the species studied. C bands are revealed by successive acid, alkaline and saline treatments at high temperature followed by staining with Giemsa. These treatments preferentially extract DNA from the regions which do not correspond to C bands leaving *in situ* the DNA of constitutive heterochromatin which is intensively stained by Giemsa (Comings et al., 1973). Acid conditions hydrolyse the glycosidic bonds between the purine bases of DNA and 2-deoxyribose, alkaline conditions denature the DNA and saline conditions at high temperature break the DNA into small fragments, which are therefore extracted (Holmquist, 1977). The higher resistance to extraction of DNA localised in C bands has not been explained. It may be related to the presence of non histone proteins closely bound to heterochromatin, and to rapid renaturation of the highly repetitive DNA sequences which C bands contain.

Protocol
The technique described here is that of Sumner (1972), slightly modified. The slides are incubated in 0.2 N hydrochloric acid for 1 hour at room temperature, followed by a rinse with distilled water. They are then incubated in a 0.3 N Ba(OH)$_2$ solution (appendix S1) at 50 °C for 30 to 60 seconds, rinsed in distilled water and incubated in 2× SSC pH 7 (appendix B4) at 60 °C for one hour, and rinsed in water. The chromosome preparations are stained with Giemsa (paragraph I.5.1) and observed under the phase contrast microscope, using an orange filter.

Remarks
In human chromosomes (Figure II.13), all centromeric regions are stained and contain alphoid type repetitive DNA (paragraph II.1.1). The regions of secondary constriction of chromosomes 1, 9, and 16 are also banded but they contain classical repetitive DNA. The short arms of the acrocentric chromo-

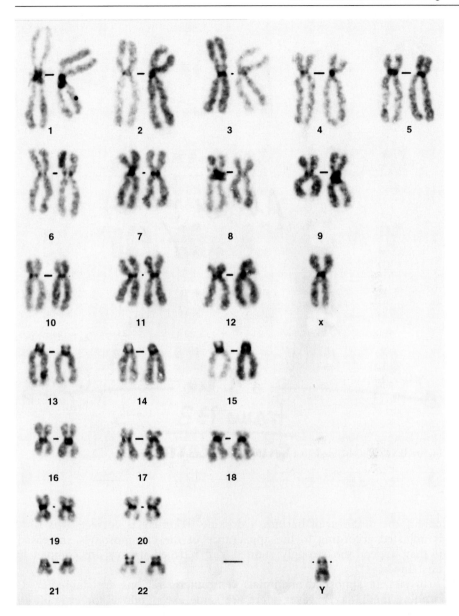

Figure II.13. CBG banded human karyotype (from Dutrillaux, 1975b, reproduced by permission of Expansion Scientifique Française).

somes are also stained. Their composition is complex since they contain several types of repetitive DNA.

In bovidae chromosomes, for example those of goat (Figure II.4), only the centromeric regions are stained and the chromatids appear uniformly

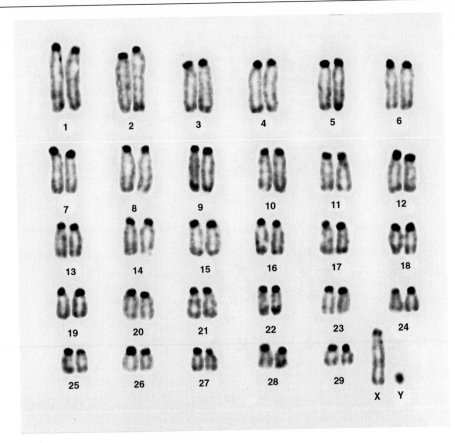

Figure II.14. CBG banded goat karyotype (photograph H. Hayes).

pale if the duration of barium hydroxide treatment is correct. This time is adjusted according to the appearance of the chromosomes, increased if they appear too strongly stained and reduced if they are completely unstained.

In fact, in almost all mammals centromeric regions are stained by C banding methods. However, there are some exceptions as for example in certain species of Prosimians of the Lorisidae family, *Nycticebus coucang* and *Perodicticus potto* (Dutrillaux et al., 1979a) or, in the rodent *Eliomys querci-nus* (Dutrillaux et al., 1979b) in which no C banding is observed in cen-tromeric regions. In some other species, only a few chromosomes display this type of banding. In the case of a robertsonian rearrangement by fusion and translocation of the centromeric regions of acrocentric chromosomes, it has been observed in several groups of species that the intensity of the C bands in the metacentric chromosomes produced by the fusion of two acrocentric chromosomes is diminished. This diminution seems to be progressive and may date the occurence of the rearrangement.

In some species, C banded constitutive heterochromatin is abundant. It is not necessarily juxta-centromeric. Thus, Plathyrhini Primates of the genus *Cebus* are characterised by the presence of long intercalated or distal segments of this type on a dozen of their chromosomes. Other Primates, like *Lemur coronatus* have very long heterochromatin segments flanking the centromeric region. Variations in heterochromatin size and composition occur quite rapidly compared to the rate of karyotype evolution, that is to say within genera, species and even between the populations of a single species. Their existence complicates the reconstitution of phylogenies.

II.5.2
CT bands

Principle
This technique, introduced by Scheres in 1976, reveals simultaneously, on human chromosomes, C bands and the T bands described earlier. It is in fact a modification of the C band technique, which itself tends to reveal T bands. Although to our knowledge the mechanism of CT banding has not yet been explored, it is evident that this method reveals the structures in chromosomal DNA most resistant to denaturation.

Protocol
Slides are incubated in a saturated solution (0.3 N) of $Ba(OH)_2$ (appendix S1) for 10 minutes at 60 °C, rinsed in distilled water and then incubated in 2× SSC pH 7 (appendix B4) for 30 minutes at 60 °C. They are then rinsed in distilled water and finally stained for 10 minutes in a solution of 0.005% Stains-All (4,5,4',5'-dibenzo-3.3'-diethyl-9-methyl-thiocarbocyanine-bromide; Serva) in a mixture of 1 vol. formamide and 1 vol. distilled water, rinsed, dried and observed, preferably under the phase contrast microscope, with an orange filter.

II.5.3
G11 bands

Principle
G11 banding is so called because it is produced by staining chromosome preparations with Giemsa at a pH of about 11. Two similar methods were introduced almost simultaneously in 1972 by Bobrow and colleagues and by Gagné and Laberge.

Protocol
In the technique of Bobrow and colleagues (1972), slides are stained for 8 minutes in a 2% solution of Giemsa, the pH of which is adjusted to 11 by addition of 0.1 M NaOH (appendix SS12).

In that of Gagné and Laberge (1972), slides are also stained in a 2% solution of Giemsa, but the solution is prepared differently. It consists of a solution of Na_2HPO_4. $12H_2O$, 0.1% (weight/vol.), Giemsa stain, 2% (vol./vol.) in distilled water to a final volume of 100 ml, the pH of which is adjusted to 11.6 by addition of 0.1 M NaOH (appendix SS12). Slides are incubated in this solution for 5 minutes and this incubation may be repeated if necessary.

Alkaline solutions of Giemsa are unstable and must be prepared extemporaneously.

Remarks

These banding techniques were first described, as specific for the secondary constrictions of human chromosome 9. They were later observed to stain other chromosome structures all of which are juxtacentromeric (Figure II.15) in man and in Pongidae species (gorilla, chimpanzee, orang-utang), but not in Hylobatidae and Cercopithecidae (Dutrillaux, 1979). In situ hybridisation using different probes of repetitive DNA has shown that the regions stained by this technique all contain classical satellite DNA with a repeated CCATT sequence element.

In man, the chromosomes carrying these structures are the metacentric or submetacentric chromosomes 1, 5, 7, 9, 10, 17, 19 and Y, and all the acrocentric chromosomes 13, 14, 15, 21, and 22. Chromosomes 1, 5, 9, 10, and 19 are stained on the proximal part of the long arm, chromosomes 7 and 17 on the proximal part of the short arm and the acrocentric chromosomes on the short arm. Chromosome Y is strongly stained on the long arm. Simultaneous in situ hybridisation with probes for alphoid and classical satellite DNA sequences each carrying a different label shows that the latter are always more centromeric (Kokalj-Vokac et al., 1993). Centromeres are, therefore, included in the alphoid sequences, but not in the classical satellite sequences. Consequently, G11 banding is juxtacentromeric but not centromeric.

G11 banding techniques give rather irregular results and often only a few of the metaphase spreads on a slide are suitably stained. The whole preparation should therefore be examined, before concluding that the experiment has failed.

II.5.4
DA-DAPI banding

Principle

DA-DAPI staining technique uses a combination of the fluororochrome 4',6-diamidino-2-phenylindole (DAPI) and the non-fluorescent peptide antibiotic distamycin A (DA). Both compounds bind to AT rich DNA. Chromosomes stained with DAPI show a fluorescent banding pattern similar to that of Q bands but less contrasted. If the chromosomes are exposed to DA before staining with DAPI, certain chromosome segments fluoresce much more brightly than others. They are situated in heterochromatic regions (Schweizer et al., 1978). In fact, only a part of the constitutive heterochromatin situated

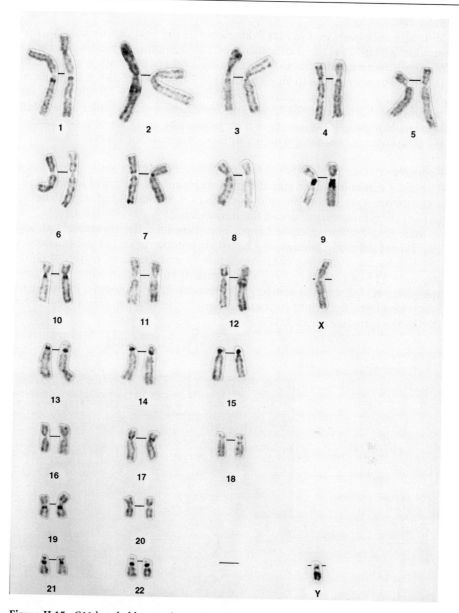

Figure II.15. G11 banded human karyotype (from Dutrillaux, 1975b, reproduced by permission of the Expansion Scientifique Française).

mainly on human chromosomes 1, 9, 16 and Y and to a lesser degree on other chromosomes is strongly stained. The sites revealed by DA-DAPI staining are identical to those revealed by G11 banding (see paragraph II.5.3) and by the antimethylcytosine antibodies (see paragraph II.2.6.2).

Protocol

Slides are incubated first in MacIlvaine's buffer pH 7 (appendix B6) containing distamycin A-HCl at a concentration of 0.2 mg/ml (appendix S3) for 15 minutes, then rinsed with the same buffer and stained with a solution of DAPI (0.4 μg/ml) in MacIlvaine's buffer, pH 7 (appendix S2) for 15 minutes. After rinsing in the same buffer at pH 7 and again at pH 7.5, the slides are mounted in this buffer at pH 7.5 with a coverslip, and excess buffer is removed by pressing between two sheets of filter paper. The slides are observed under the fluorescence microscope with a filter appropriate for DAPI.

Remarks

The main difficulty encountered in this technique is the rapid loss of fluorescence intensity during ultraviolet irradiation. Storage of slides for 15 hours in the refrigerator may stabilise the fluorescence.

After DA-DAPI staining of human chromosomes, the chromatids generally fluoresce weakly, whereas the heterochromatic regions of chromosomes 1, 9 and 16, the short arm of chromosome 15 and the distal region of chromosome Y fluoresce very brightly (Figure II.16). Certain other juxtacentromeric regions are also relatively fluorescent. After prolonged exposure to ultraviolet light, the bright regions fade and Q banding patterns appear to be due to the fluorescence of DAPI.

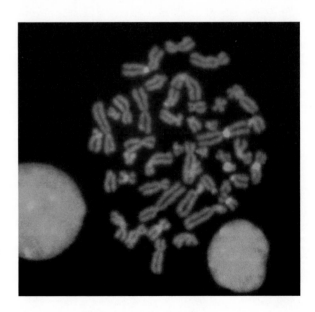

Figure II.16. Human chromosome spread after DA-DAPI staining (photograph C. Bourgeois).

II.6
Staining of nucleolar organiser regions NOR

H. Hayes and B. Dutrillaux

The nucleolar organiser region or NOR is the chromosomal site which contains the genes for ribosomal RNA and at which the nucleolus is formed. The NORs can be revealed on chromosomes after a treatment in alkaline conditions which causes extraction of the DNA, RNA and histones, followed by silver nitrate staining which reacts with the proteins specifically associated with the NORs. Only those nucleolar organiser regions which have participated in the formation of nucleoli during the interphase preceding cell harvest are revealed by silver nitrate. This indicates that the proteins stained by silver nitrate are those involved in the transcription of ribosomal RNA genes (Miller et al., 1976). Many variants of the original technique of Goodpasture and Bloom (1975) exist. The protocol described here is a modification of that of Lau and Arrighi (1977).

Protocol
The slides are incubated in borate buffer pH 9.2 (appendix B7) for 30 minutes at room temperature, rinsed in distilled water, and airdried. They are then mounted in a 50% solution of silver nitrate (appendix S4) with a coverslip, placed in a humid chamber and incubated in a water bath at 65 °C for 1 hour. After rinsing with distilled water, the slides are stained for 1 minute with 1% Giemsa (see paragraph I.5.1) and observed under the microscope in normal light.

Remarks
Accumulation of silver grains are observed on chromosomes carrying NORs (Figure II.17). Chromosome identification is carried out by a sequential method using either Q bands revealed before silver nitrate staining, or RBA bands revealed after silver nitrate staining. The number of chromosomes carrying NORs is specific for each species but the number of NORs observed after staining is variable because it depends on the transcriptional activity of ribosomal RNA genes during interphase.

II.7
Techniques for sequential banding

B. Dutrillaux

Principle
Sequential banding methods permit the successive revelation of two or more banding patterns on the same preparation and thus their direct comparison. This procedure is very useful for the following purposes:

Figure II.17.
Chromosome spread of
Chinese Meishan pig
showing the four
chromosomes carrying
nucleolar organising
regions (NOR)
(photograph P. Popescu).

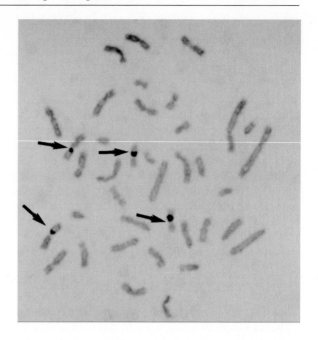

- a precise determination of the equivalences between for example Q and R banded chromosomes of the same species especially in the case of species such as cattle and goat which contain many acrocentric chromosomes with similar morphology that are difficult to identify,
- the localization of a breakpoint with respect to a banding pattern,
- the identification of chromosomes carrying NORs or containing constitutive heterochromatin.

In these cases it is necessary to use complementary banding methods and this requires precise ordering of the treatments. The principle of this ordering is very simple: the treatments should be applied progressively from the least aggressive towards the most aggressive. For example, simple stains such as Giemsa or fluorescent intercalating agents like quinacrine mustard do not attack the structure of the chromosome and should be used for the first banding method. However, it should be remembered that exposure to ultraviolet light required for the observation of chromosomes stained by a fluorochrome will alter their structure, in particular by introducing numerous breaks in their DNA which may then be released into the mounting medium. The irradiation time in ultraviolet light should therefore be limited as much as possible, especially if it is to be followed by a banding technique involving a denaturation treatment.

Many protocols for sequential banding have been described, each laboratory having worked out the combination of techniques best adapted to its needs. One example suitable for precise identification of a chromosome, a

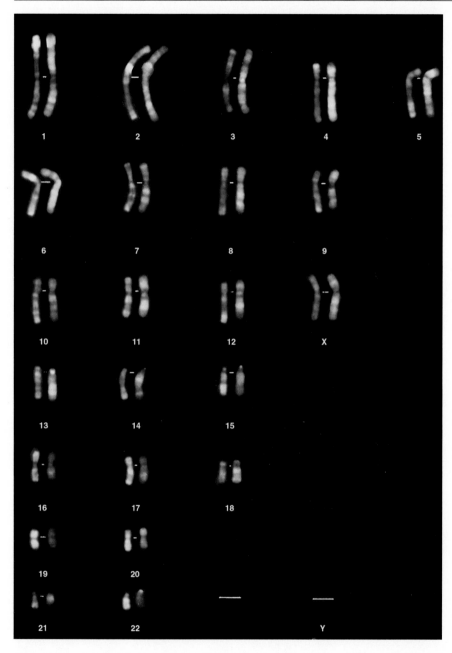

Figure II.18. Sequential Q and R banding on human chromosomes (from Dutrillaux, 1975b, reproduced by permission of the Expansion Scientifique Française).

chromosome region, or a band is presented here. To obtain several successive banding patterns on the same slide, for example: G followed by C or Q followed by R and then C, or R followed by NOR staining, two factors should be considered. The first is the duration of the different treatments. It is often necessary to reduce that of later steps in the sequence especially if earlier steps have caused partial denaturation of DNA. The second is the type of destaining technique applied between the banding treatments which should be adapted according to the following banding procedures.

Protocol for sequential Q, R and C banding

Q and R banding permit analysis of euchromatin and C banding reveals the constitutive heterochromatin regions. The chromosome preparations are first stained with quinacrine mustard (see paragraph II.2.1) and Q banding patterns are revealed. Interesting mitoses are photographed and their positions are recorded. The coverslip is carefully removed and the slides are incubated at 87 °C for 10 to 20 minutes in a 1× bicarbonate free Earle's solution pH 6.5 (appendix B9) to reveal RHG bands (see paragraph II.2.4). This treatment involving only one incubation for a shorter time in Earle's solution is less severe than that previously described (see paragraph I.5.3) but produces exploitable R band patterns. The slides are then stained with Giemsa (paragraph I.5.1) or with acridine orange (paragraph I.5.3) and the same mitoses are photographed again (Figure II.18). The slides are then washed quickly in a series of toluene baths to remove the immersion oil. The treatment for C bands described in the paragraph II.5.1 is applied directly without destaining, but the duration of incubation in Ba(OH)$_2$ is reduced by half.

Protocol of sequential Q and NOR banding, or R and NOR banding

The method for staining nucleolar organiser regions (NOR) described in the paragraph II.6 can be applied after Q or R banding. Slides should first be freed from immersion oil by rinsing in toluene, then destained in two successive baths of ethanol at 70% and 50% and finally fixed by incubation for 5 minutes in Carnoy's solution (6 vol. ethanol, 3 vol. chloroform, 1 vol. acetic acid). The resulting acidification of the slides improves NOR staining. It should be noted that it is in general difficult to obtain satisfactory NOR staining after the enzymatic digestion treatments involved in GTG banding.

III
In Situ Hybridisation Techniques

H. Hayes and B. Dutrillaux

III.1
Introduction

In situ hybridisation permits the visualisation of the hybridisation of specific nucleotide sequences (probes) on chromosomes or cell or tissue slices. It is used for the physical mapping of chromosomes, for chromosome rearrangement analysis, for comparative gene mapping between different species and also for studies of chromosome structure and evolution. For genome mapping, it is the most direct and precise method for the localisation of a gene or DNA sequence on a chromosome.

Its successful use requires the detection of the signal carried by the hybridised probe and the identification of the chromosome(s) to which it is bound. It also depends on the formation of a DNA/DNA or DNA/RNA hybrid, which is possible only if the two interacting strands are totally or partially complementary and if the two parental molecules are denatured before hybridisation. In addition, detection of the hybrid under the microscope requires that the nucleotide sequence hybridised to the chromosomal DNA be labelled. The technique was described first in 1969 by two different groups (Pardue and Gall, 1969; John et al., 1969). At that time, the only technique sensitive enough to detect hybridised sequences was autoradiography and therefore the use of radioactively labelled probes. Because of the low sensitivity of this technique only highly repetitive or amplified sequences could then be localised. Since 1969 *in situ* hybridisation has been developed through two major phases, the first based on the use of radioactive and the second on non radioactive probes and at the same time techniques of chromosome identification have been improved.

III.1.1
In situ hybridisation using radioactive probes

Refinement of the procedures for DNA segment cloning and radioactive labelling [labelling by nick translation (Rigby et al., 1977) and random priming (Feinberg and Vogelstein, 1983)], improvement of the molecular hybridisation methods [use of formamide which permits low hybridisation temperatures (37–42 °C) and preserves chromosome morphology and of dextran sulphate which increases the number of hybridised molecules (Wahl et al., 1979)], and of chromosomal banding techniques have permitted precise localisation of unique DNA sequences on metaphase chromosomes (Harper and Saunders, 1981; Gerhard et al., 1981).

However, the use of radioactive probes (^{3}H or ^{125}I) for *in situ* hybridisation presents several difficulties:

- security rules concerning the use of radioactivity must be observed
- the exposure time for autoradiography is very long (1 to 3 weeks)
- because of a generally high background, statistical analysis of a large number of metaphases is necessary (100 to 200)
- the resolving power of signal detection is low because of the dispersion of silver grains in the photographic emulsion.

In spite of these difficulties, this technique has permitted the localisation of 1500 to 2000 genes on human chromosomes. It has also been used in studies on animal species but for a shorter time and not so intensively. For example, since 1984 genes have been localised on chromosomes of pig (Geffrotin et al., 1984; Echard et al., 1986), cattle (Fries et al., 1986), horse (Mäkinen et al., 1989; Ansari et al., 1988), sheep (Hediger, 1988; Mahdy et al., 1989) and goat (Simi et al., 1989; Chowdhary et al., 1991).

III.1.2
In situ hybridisation using non radioactive probes

The introduction of non radioactive labels to mark DNA probes led to rapid progress in the technique of *in situ* hybridisation. The wide range of methods of non radioactive labelling which have been developed fall into two groups: direct and indirect techniques (Ward, 1990; Habeebu et al., 1990; Lichter et al., 1991; Trask, 1991).

a. **Direct techniques** are based upon the incorporation into nucleic acids of nucleotides coupled to a fluorochrome (fluoroscein-11-dUTP), which can be directly visualised after the hybridisation reaction (Wiegant et al., 1991). However, this method is not sufficiently sensitive for the hybridisation with short probes or non repetitive sequences.

b. **Indirect techniques** are based upon the preparation of probes containing modified nucleotides which can be detected by antibodies coupled to a fluorochrome or by enzymatic reaction. These probes are prepared:
- either by the incorporation of nucleotides, coupled to biotin (Langer et al., 1981), to digoxygenin (Kessler, 1990) or to dinitrophenol (Langer et al., 1981),
- or by direct chemical modification of nucleotides in the probes, for example by reaction with N-acetoxy-2-acetyl-aminofluorene (Tchen et al., 1984; Landegent et al., 1984). At present, incorporation of a nucleotide-biotin complex (biotin-11-dUTP) and its detection by chemical affinity (biotin-avidine) or by immunoaffinity (anti-biotin) followed by a second reaction with an antibody coupled to a fluorochrome

(fluorescein, rhodamine) is the most sensitive and efficient technique for the localisation of unique sequences or short probes, from 0.5 to 1 kb (Lawrence et al., 1988; Pinkel et al., 1988; Viegas-Péquinot et al., 1989; Cherif et al., 1989; Lemieux et al., 1992).

In situ hybridisation with non-radioactive probes presents several advantages as compared to the use of radioactive probes:

- Since radioactivity is not used all manipulations are simplified, the only requirement is access to a fluorescence microscope,
- the hybridisation signal is visualised directly, not via a photographic emulsion; the resolving power of the technique and thus the precision of the localisation of the hybridised sequence on the chromosome are increased,
- the whole procedure is much faster (one or two days); in addition, it is possible to detect several probes simultaneously by using different labelling and immunodetection systems (Nederlof et al., 1990; Ried et al., 1992).

III.1.3
Identification of hybridised chromosomes

Both visualisation of the hybridisation signal and chromosome identification are necessary for the successful use of *in situ* hybridisation techniques. However, these techniques involve conditions which modify the behaviour of chromosomes towards banding treatments and may therefore complicate their identification.

Until 1990, *in situ* hybridisation was performed in two steps:

- detection and photography of the hybridisation signal,
- generation and photography of the chromosome banding pattern.

The two pictures were then compared to identify the chromosomes carrying the hybridisation signal and to localise the latter precisely. Chromosome bands may be revealed before or after hybridisation (Malcolm et al., 1986; Szabo, 1989).

If bands are revealed before hybridisation, the banded metaphases are identified and photographed and after hybridisation the same metaphases are photographed again to detect the hybridisation signals. In this case, a large number of metaphase spreads must be observed since not all will display hybridisation signals.

If bands are revealed after the visualisation of the hybridisation signal, the number of metaphase spreads examined can be reduced but the quality of the banding patterns decreases. In both cases two photographs are necessary.

The introduction of non radioactive probes for *in situ* hybridisation not only simplified the technique but stimulated the development of new strategies permitting the simultaneous observation under the fluorescence microscope of the chromosome banding pattern and the signal of the hybridised probe. With these methods, the number of metaphase spreads analysed is reduced and genes can be localised with sufficient precision to situate them.

Most of these strategies use the so-called "dynamic" banding method involving BrdU incorporation and propidium iodide staining. Different authors (Cherif et al., 1990; Fan et al., 1990; Takahashi et al., 1990; Lemieux et al., 1992) propose different protocols which however all lead to the same result.

III.2
Methods

The literature contains many protocols which differ according to the type of probe labelling, the conditions of denaturation of the DNA probes and chromosomes, the conditions of hybridisation and of washing after hybridisation, and the methods of chromosome identification. Only protocols for radioactive and non radioactive *in situ* hybridisation used routinely in our laboratories are described in this manual. They can be divided into the following general steps. Some of these have already been described, the others will be discussed here:

- preparation of chromosome spreads
- radioactive or non radioactive labelling of the probe
- pretreatment of the chromosome preparation with ribonuclease A
- denaturation of the chromosomal DNA
- probe denaturation
- *in situ* hybridisation
- posthybridisation washes
- detection of hybridisation signals either by autoradiography, or by immunoreaction
- revelation of RBG or RBP bands
- observation and photography of the results

Successful *in situ* hybridisation depends mainly on the quality of the chromosome preparations and probes used.

III.2.1
Preparation of chromosome spreads

Chromosome spreads are obtained from fibroblast cultures synchronised by thymidine or amethopterine and labelled with BrdU during the second half of the S phase (according to protocols described in paragraphs I.2 and II.3.1), to generate R banding patterns.

III.2.2
DNA probe labelling

III.2.2.1
Non radioactive labelling of long DNA probes (>1 kb)

Principle

DNA probes longer than 1 kb can be labelled either by nick translation (Rigby et al., 1977) or by random primed labelling (Feinberg and Vogelstein, 1983). Although the two methods achieve equally efficient probe labelling, nick translation is generally used because it yields a larger quantity of labelled DNA per experiment (500 ng to 1 µg) than random primed labelling (50 ng). *In situ* hybridisation experiments, generally, require quantities of the order of 100 ng of probe per slide.

The nick translation process utilises two enzymes: **DNase** the endonuclease activity of which creates single strand nicks in double stranded DNA and **DNA polymerase I** which binds to the nicked sites, removes nucleotides from the 5' end of the nicked strand by means of its 5'–3' exonuclease activity, and adds nucleotides to the 3' end through its polymerase activity thus causing the nick to move along the DNA strand, hence the name nick translation. If one or more of the incorporated nucleotides are coupled to biotin, labelled DNA is obtained. For *in situ* hybridisation experiments it is not necessary to excise the DNA fragments inserted in the cloning vector. The process is carried out using commercial nick translation labelling kits. The length of the probes, labeled according to the kit supplier's instructions, should be greater than about 2–3 kb to obtain a detectable signal. The biotinylated probes can be stored for several months at −20 °C.

Protocol for nick translation labelling

500 ng of DNA are incubated for 1 hour at 15 °C in presence of the different dNTP with 11-dUTP-biotin (Sigma B6780) at a final concentration of 2.10^{-2} or 5.10^{-2} mM or with 14-dATP-biotin (BioNick kit from Gibco-BRL), DNase, DNA polymerase I, and buffer according to the instructions of the supplier of the labelling kit (Boehringer Mannheim or Gibco-BRL). The reaction is stopped by the addition of EDTA, the labelled probe is separated from free nucleotides by passage through a G50 Sephadex column and then is precipitated in the presence of 100 µg of sonicated salmon or herring sperm DNA and, if necessary, of a variable quantity of sonicated total genomic DNA (125–500 µg) of a species homologous to the probe (appendix M4). The DNA is resuspended in 36 µl of hybridisation mixture (appendix H2) and then diluted to a concentration suitable for *in situ* hybridisation.

III.2.2.2
Non radioactive labelling of short DNA probes (0.25–1.5 kb)

Principle
The intensity of the hybridisation signal emitted by short probes (<1 kb), labelled non radioactively by nick translation and used for *in situ* hybridisation is too low to be detected. In some cases this difficulty can be overcome by using the technique of enzymatic amplification (Polymerase Chain Reaction or PCR) and labelling of the probe in the presence of a biotinylated nucleotide (Richard et al., 1994). This procedure requires the addition of two primers to the ends of the DNA fragment to be amplified and labelled and gives satisfactory results only with probes less than 1.5 kb long.

Protocol
The different components of the PCR reaction mixture are mixed in a final volume of 50 μl:

- 200 ng of genomic DNA of the species homologous to that of the sequence to be amplified or 0.2 ng of the DNA of the plasmid containing the sequence to be amplified
- 50 pmole of each primer
- 200 μM dATP, 200 μM dGTP, 200 μM dCTP, 200 μM biotin-dUTP/dTTP (3/1) (biotin-11-dUTP, Sigma or biotin-16-dUTP, Boehringer Mannheim or biotin-14-dATP, Gibco-BRL)
- buffer, MgCl$_2$, Taq DNA polymerase according to the supplier's instructions
- distilled water to a final volume of 50 μl.

Amplification and labelling is then achieved by a series of repetitive temperature cycles adapted to the sequence to be amplified. The labelled probe is separated from free nucleotides by passage through a G50 Sephadex column, as in the nick translation method. The incorporation of the biotinylated nucleotide in the amplified product is verified by comparing its apparent size, after migration in a 1.5% agarose gel with that of the same product prepared without biotin which migrates more rapidly than the biotinylated fragment.

The rest of the preparation procedure (addition of competitor DNA, precipitation, resuspension in the hybridisation medium mix) is the same as in the previous paragraph.

Remarks
All these methods of labelling can also be carried out with a nucleotide coupled to digoxygenin (digoxygenin-11-dUTP, Boehringer Mannheim) in place of the biotinylated nucleotide. Probes labelled in this way give less easily detectable fluorescence signals which limits the use of digoxygenin-11-dUTP to the localisation of repetitive DNA sequences or to that of unique DNA sequences >5 kb long.

III.2.2.3
Radioactive labelling

Radioactive labelling of probes using nick translation or short random DNA primers follows the same principles as the non radioactive labelling methods described above. The biotinylated nucleotide is replaced by a mixture of three ^{3}H labelled nucleotides: ^{3}H-dCTP, ^{3}H-dATP and ^{3}H-dTTP. The ^{3}H-dGTP is not used as it is unstable and only weakly radioactive.

Protocol for the use of the nick translation method
500 ng of DNA are incubated for 2 hours at 15 °C in the presence of 30 µCi of each of the tritiated nucleotides (^{3}H-dCTP, ^{3}H-dATP and ^{3}H-dTTP), dGTP, DNase, DNA polymerase I and buffer, according to the instructions of the supplier of the labelling kit (Boehringer Mannheim). The rest of the preparation procedure is the same as that for non radioactive labelling. The specific radioactivity of the probe is determined. It should be of the order of 10^8 cpm/µg DNA. Radioactive probes can be stored for 3 months at −80 °C. This protocol can be used for DNA sequences varying in length from 0.5 to 10 kb.

III.2.3
Pretreatment of the chromosome preparations using ribonuclease A

Principle
Chromosome preparations are treated with ribonuclease A, to remove endogenous RNA which can be a source of background.

Protocol
Slides are incubated in 2× SSC buffer pH 7 (appendix B4) containing 100 µg/ml of ribonuclease A (10 mg/ml stock solution diluted 100 fold, appendix H1) for 1 hour at 37 °C, rinsed once in 2× SSC buffer for 10 minutes at room temperature, dehydrated by three successive 10 minute washes at room temperature in 50%, 75% and 100% ethanol and finally airdried.

III.2.4
Chromosomal DNA denaturation

Principle
Chromosomal DNA must be denatured for hybridisation of the DNA probe to occur. Classical DNA denaturation treatments (high pH or high temperature) alter the morphology of chromosomes and prevent their identification and the localisation of the hybridisation signal. It is therefore necessary to use milder conditions which cause a sufficient degree of denaturation of chromosomal DNA without altering chromosomal morphology. Most laboratories in which *in situ* hybridisation on chromosomes is carried out use a

70% solution of formamide at pH 7 to denature chromosomal DNA since the chemical characteristics of formamide decrease the temperature at which DNA denatures.

Protocol

Slides are incubated for 2 minutes at 70 °C in preheated 2× SSC buffer pH 7 (appendix B4) containing 70% (vol./vol.) of deionised formamide (appendix M7). The slides are treated one by one, allowing an interval of several minutes between each treatment in order to maintain the temperature. They are then rinsed for 2 minutes in each of three successive baths of 2× SSC, 2× SSC and 0.1× SSC buffer pH 7 (appendix B4) kept on ice and are dehydrated by four successive 2 minute washes in ice cold 50%, 75%, 100% and 100% ethanol. They are stored in the last 100% ethanol bath and dried quickly just before addition of the probe.

III.2.5
Probe preparation, labelling and denaturation

After precipitation, probes are resuspended in the hybridisation medium (appendix H2) and diluted to a suitable concentration with the same medium in a final volume of 10 to 20 µl. Non-radioactive probes are generally hybridised at a concentration of 6.5 ng/µl, but the concentration can be diluted up to 100 fold. The optimal concentration of radioactive probes depends on the specific activity of the radioactivity and should be determined by successive approximations (usually about 10^6 cpm/slide). Probes are denatured at 100 °C for 10 minutes and then placed on ice for 10 minutes, before being deposited on the slides.

III.2.6
In situ hybridisation

10 to 20 µl of hybridisation medium containing the denatured probe are placed on the cell spread, which has been localised in advance, and are covered with a piece of plastic film of the type sold for domestic cooking and freezing which adheres well to the slide and does not damage the chromosome preparation. The slides are incubated at 37 °C for about 20 hours in a medium saturated by 2× SSC buffer pH 7 (appendix B4) containing 50% formamide.

III.2.7
Posthybridisation washes

III.2.7.1
Radioactive probes

Much higher backgrounds are observed with radioactive probes than with non-radioactive probes and therefore when the former are used the number and duration of rinse baths must be increased. In this case, after removal of coverslips, slides are washed for 10 minutes in each of the following series of 200 ml baths:

- 3 successive baths containing 50% deionised formamide (appendix M7) in 2× SSC buffer pH 7 (appendix B4) at 39 °C
- 3 successive baths in 2× SSC buffer pH 7 at 39 °C
- 3 successive baths in 2× SSC buffer pH 7 at room temperature.

They are then placed in a 1 litre bath of 0.1× SSC buffer pH 7 at room temperature for 30 to 60 minutes, and transferred to a second 1 litre bath of 0.1× SSC buffer pH 7 at 4 °C for 30 to 60 minutes and finally dehydrated through four successive 10 minute baths in ethanol 50%, 75%, 100% and 100%.

III.2.7.2
Non-radioactive probes

After hybridisation, the coverslips are carefully removed and the slides are washed for 2 or 3 minutes in each of the four following baths at 38 °C to eliminate unhybridised or non specifically hybridised DNA, two baths in 2× SSC buffer pH 7 (appendix B4) containing 50% formamide, two baths in 2× SSC buffer pH 7 or one in this buffer followed by one in 1× SSC pH 7. The washing conditions (concentration of the washing solutions, duration and temperature) can be adjusted according to the length and constitution of the hybridised DNA probe. Those described here are suitable for probes between 2 and 15 kb long and localisation of unique DNA sequences.

If preparations present a significant background, it is rarely possible to remove this by modifying only the slide washing conditions. It is necessary to verify the quality of the probe preparations, of the slides and of the washing solutions. If the background seems to be due to the quality of the probe preparations, different concentrations of the DNA competitor (in the case of DNA fragments containing repetitive sequences) or different decreasing concentrations of probe must be tested.

III.2.8
Hybridisation signal detection and banding

III.2.8.1
Radioactive probes: autoradiographic detection

Principle
Radioactive probes are detected by autoradiography. The slides are covered with a thin coat of photographic emulsion and allowed to dry. The *beta* particles emitted by the tritium interact with the emulsion and produce the silver grains which after revelation and fixation are visible under the optical microscope.

Chromosome spreads presenting hybridisation signals are localised and photographed, and the slides are then treated by the FPG banding method (see paragraph II.3.1.1) to reveal RBG bands. The same chromosome spreads are photographed again to localise chromosomes carrying the hybridisation signal.

Protocol

■ **Autoradiography.** This operation must be carried out in a dark room in total darkness (tell tale lights on switches etc should be masked) and all accessories must be prepared beforehand: waterbath at 42 °C, borel tube, 25 and 50 ml test tubes, 25 ml beaker, clean blank slides, drying rack, black plastic dishes containing bags of desiccant, distilled water containing 2% of glycerol, Kodak emulsion NTB2. The emulsion must be manipulated with great care to avoid exposure which causes loss of experimental results.

A 25 ml beaker is filled with photographic emulsion using a spatula and placed in a waterbath at 42 °C to melt the emulsion. Distilled water containing 2% of glycerol is also preheated to 42 °C. Using a measuring cylinder carrying markers at 15 and 30 ml, 15 ml of melted emulsion is diluted with 15 ml of aqueous 2% glycerol. The diluted emulsion is poured gently, avoiding bubbles, into a borel tube covered by aluminium foil which is then placed in the water-bath at 42 °C. To verify that the diluted emulsion is completely dissolved and free from bubbles a clean blank slide is dipped into it, withdrawn and examined in infrared light. The presence of small visible craters in the emulsion layer reveals the existence of bubbles. When the emulsion is ready for use, the hybridisation slides are dipped into it one by one, for 1 minute, drained vertically for a few minutes on filter paper and then held in darkness in a vertical position for 2 to 3 hours to yield a sufficiently dry and thin emulsion layer. Finally, the slides are placed in a black box containing desiccant which is then sealed with black electric tape. The box is stored at 4 °C for the required exposure time (8 to 15 days). The optimal time of exposure is determined by developing a slide every three days.

■ **Slide development.** In the dark room the slides are removed from the refrigerator and left for at least 2 hours to attain room temperature (~20 °C). They are then dipped successively for 1 minute into baths at 20 °C containing developer (Dekta or Kodak D19) and fixative (Ilford) followed by two baths of distilled water, drained in a rack and dried for 24 hours. They are then stained in Sorensen's buffer pH 6.8 (appendix B3) containing 1.5 or 4% Giemsa stain for 1 to 5 minutes according to the degree of staining desired. The preparations are then examined under the microscope in normal light or phase contrast. Chromosome spreads presenting a hybridisation signal are localised and photographed with Kodak technical pan film, at the previously determined optimal sensitivity for the type of microscope used. 100 to 150 chromosome spreads should be photographed.

■ **RBG banding.** After degreasing with methylcyclohexane, slides are treated according to the FPG banding method (see paragraph II.3.1.1 for RBG banding). The success of this operation requires careful control of the method because the morphology of the chromosomes is considerably altered by the treatments to which they are subjected. It is often necessary to reduce the duration of the denaturation treatment at 87 °C or even to omit this treatment. After staining, chromosome spreads previously shown to carry signals are photographed again. The RBG bands are much paler and less well defined than on non hybridised slides, but chromosome identification is possible.

III.2.8.2
Non-radioactive probes: immunoreaction detection

Immunodetection of the hybridisation signals on chromosomes is performed in four steps: saturation of nonspecific antibody binding sites, reaction with the first anti-biotin antibody, reaction with a second fluorescein labeled antibody directed against the first antibody, chromosome counterstaining with propidium iodide.

Protocol
Slides are incubated for 10 minutes in each of successive baths of PBT solution at room temperature (appendix H3). 50 to 100 µl of anti-biotin antibody solution (appendix H4) are then dropped on the slides, coverslips are added, and the slides are incubated for 1 hour at 37 °C in a humid chamber protected from light. After removal of coverslips and two washes in PBT solution, 50 to 100 µl of fluorescein labeled anti-goat IgG antibody (appendix H5) are dropped onto the slides which are again covered and incubated as described above. A few minutes before the end of the incubation with the second antibody, 100 µl of a solution of propidium iodide at a final concentration of 1 ng/µl (appendix S12) are placed on the slides and incubation is continued for a further 6 minutes at room temperature in the dark, after which the slides are rinsed in phosphate buffer (appendix B10), mounted in the same solution with a coverslip, and stored at 4 °C until examination.

III.2.9
Microscopy and photography

III.2.9.1
Radioactive probes

(see paragraph III.2.8 on hybridisation signal detection and banding)

II.2.9.2
Non radioactive probes

Chromosome preparations are observed under the fluorescence microscope in the presence of PPD11 (see paragraph II.3.1). The microscope is equipped with a 100 W mercury-vapour lamp for fluorescence excitation and with a Leitz I2/3 filter set (transmission interval 450–490 nm) which permits observation of the light emitted by the fluorescent dyes, yellow green fluorescence emitted by fluorescein and red orange emitted by propidium iodide. Thus, the hybridisation signals (yellow green) and the RBP chromosome bands (red and dark) can be observed simultaneously, and hybridised DNA fragments can be mapped precisely on chromosomes (Figures III.1a and III.1b). The photographs are taken with a Kodak ektachrome 400 ASA film, using a sensitivity of 320, 400, or 500 ASA according to signal intensity. The best prints are enlarged with a slide duplicator and Kodak ektachrome SE duplicating film for colour prints, or Ilford PAN F 50 ASA film for black and white prints.

III.3
Remarks on other applications of *in situ* hybridisation

Recently, new methods which enlarge the field of application of *in situ* hybridisation have been developed. Competitive in situ hybridisation consists in the hybridisation of a mixture containing the probe and a large excess of competitor DNA, in order to avoid the non specific signals produced by hybridisation of the repetitive segments present in most genomic DNA probes and dispersed in all chromosomes (Landegent et al., 1987; Lichter et al., 1990). By this method a wide range of fragments containing repeated sequences isolated from libraries of genomic DNA can be mapped by hybridisation. The chromosome painting method permits the revelation of a whole chromosome by using as the probe the total DNA of a library specific for a single chromosome isolated by cytofluorometric sorting (Lichter et al., 1988a). This method has facilitated the detection of numerical or structural chromosome rearrangements and the identification of chromosomal markers in the cells of patients with genetic diseases (Pinkel et al., 1988; Lichter et al., 1988b) or in tumour cells (Cremer et al., 1988). More recently, this method has also been used to identify chromosomes isolated by cytofluorometric sorting from pig (Yerle et al., 1993), domestic cattle (Schmitz et al., 1995) and mouse (Breneman et al., 1995; Rabbitts et al., 1995). Chromosome specific DNA libraries are now commercially available for all human chromosomes. They may be used for cross hybridisation to chromosome spreads of species other than man. This technique of heterologous chromosome painting (ZOO-FISH) permits the identification of chromosomal segments conserved among species (Figure III.2). It produces in a direct and simple procedure an overall picture of the number and size of segments conserved between human chromosomes and those of another species and thus of these chromosomal homologies (Wienberg et al., 1992; Jauch et al., 1992; Scherthan et al., 1994; Rettenberger et al., 1995a; Stanyon et al., 1995; Hayes et al., 1995; Rettenberger et al., 1995b; Apiou et al., 1996). Inversely, painting of human chromosomes by pig chromosome specific DNA librairies, has permitted more precise definition of homologies between human and pig chromosomes (Goureau et al., 1996). Comparative genomic DNA hybridisation (CGH) is used to search for variations in the genetic content of a genome without direct analysis of the corresponding karyotype or use of specific

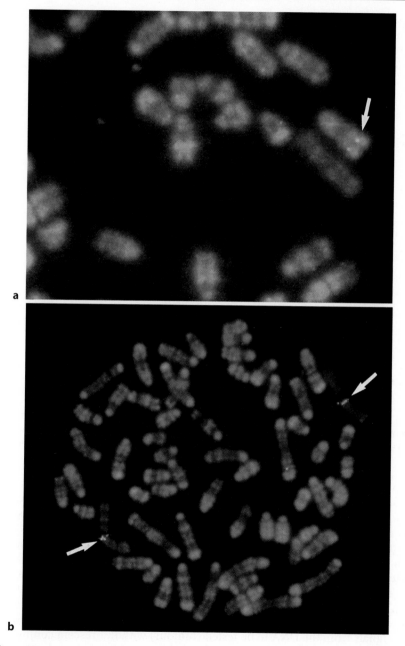

Figure III.1. *In situ* hybridisation of a biotinylated probe on RBP banded cattle chromosomes (photograph H. Hayes).
(a) GGTA1 gene sequence on cattle chromosome 11q26, the sequence of 0.94 kb was labelled by PCR with biotine-11-dUTP.
(b) A 30 kb sequence specific of cattle chromosome X inserted in a cosmid and labelled by nick translation in the presence of biotine-11-dUTP.

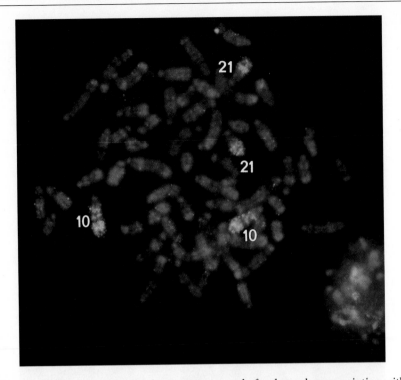

Figure III.2. RBP banded cattle chromosome spread after heterologous painting with a biotinylated DNA library specific for human chromosome 14 (Cambio Chromosome Painting System). Cattle chromosomes 10 are painted on three different segments, two small segments in the proximal half and a segment covering about the distal half of the chromosome and cattle chromosomes 21 are painted on their distal half (photograph H. Hayes). These four segments represent the regions conserved between human chromosome 14 and cattle chromosomes 10 and 21 (Hayes, 1995).

probes (eg in the genetic complement of tumour cells in which parts of chromosomes or whole chromosomes are often observed to be in excess or missing). In this recently introduced method (Kallioniemi et al., 1992), a mixture of reference genomic DNA prepared from normal cells and labelled with dUTP-digoxygenin and the genomic DNA to be tested prepared from tumour cells and labelled with dUTP-biotin is hybridised to chromosome spreads from normal cells in the presence of a large excess of competitor DNA. Hybridisation signals from reference DNA (normal) are detected with anti-digoxygenin-rhodamine antibodies (red fluorescence) and hybridisation signals from the DNA to be tested (tumour) with fluorescein conjugated avidin (green fluorescence). The relative amounts of tumour and reference DNA sequences which hybridise to the chromosomal DNA are dependent on the number of the copies of these sequences in the two DNA samples and can be evaluated by measuring the ratio of the green and red fluorescences. Thus, if the DNA to be tested (tumour) comes from cells with a chromosome in excess, the ratio of green to red fluorescence is increased by a factor of 1.5 whereas if this DNA comes from cells with a missing chromosome, this ratio is decreased by a factor of 0.5. The CGH method which is still experimental, should have applications in medicine (detection of chromosomal imbalances) and in fundamental research (studies of the differences between the genomes of closely related species or between those of individuals of the same species (du Manoir et al., 1993).

IV
Methods of Germ Cells Study

IV.1
Meiosis in male

P. POPESCU

In Mammalian male the spermatogenetic cycle is a process which assures the spermatozoids production. This cycle involves three steps:

- division of spermatogonia strains which allows the perenniallity of the spermatogenesis.
- spermatogenesis, which includes meiosis. At the end of meiosis the chromosome complement is haploid.
- spermiogenesis which transforms the spermatids into spermatozoa.

During spermatogenesis one round of chromosome duplication, but two rounds of cell division are produced (first and second meiotic division) leading to gametes formation. Each of them contains a haploid chromosome complement. The first meiotic division begins with a stage of prophase much longer than in mitotic division. This stage involves several steps: *leptotene* (chromosome individualisation), zygotene (pairing of homologous chromosomes), pachytene (genetic recombination stage in synaptonemal complexes), diloptene (separation of two pairs of sister chromatids except in chiasmas) and diacinese (chromatid contraction). Different troubles can be produced during meiosis, that is, the theoretical efficiency of four spermatids for a first class spermatocyte is not always attained.

Meiosis study is useful for a better understanding of the origin of male sterility and the behaviour and the segregation of abnormal chromosomes. This kind of study is performed on a testicular biopsy by surgical means.

IV.1.1
Classical method

Protocol
The fragment of testicular biopsy is placed in a Na$_3$citrate. 2H$_2$O solution (2, 24% weight/vol.) at room temperature and lightly dilacerated with a forceps and a needle. The tissue fragments are then transferred in this solu-

tion (1% weight/vol.) and dilacerated. The cell suspension is then pipetted and transferred to a conical tube. A solution of 1% Na$_3$citrate. 2H$_2$O is added, the fragments are again dilacerated and the supernatant is discarded. This operation can be repeated several times, as long as the suspension becomes opalescent through the presence of the cell suspension issued from seminiferous tubes. The cells are immersed in hypotonic solution for about 30 minutes. The conic tube is then centrifuged at 400 g for 5 minutes.

The fixation is identical to paragraph I.4. The spreads are prepared either on dried slides (allow three drops of cell suspension to fall on the slide and blow when Newton rings appear) or on slides covered by a thin aqueous film (allow one drop of cell suspension to fall and air dry).

The preparations are stained using Giemsa (according to paragraph I.5.1) or toluidine blue (aqueous solution 0.5%). This method, introduced for the first time by Evans et al. (1974), allows the observation of the cells in diacinese and metaphase of the first and the second meiotic division (Figures IV.1, a, b, c).

Different alternatives have been proposed since the original method description. Therefore, Luciani et al. (1975) recommend modifications which improve chromosome dispersion in pachytene and their characteristic structures become visible. Fragments of biopsy are directly immersed in a hypotonic KCl solution of 0.44% (weight/vol.) for 8–10 hours at room temperature. They are then fixed for 12–18 hours in a mixture (3 vol. methanol – 1 vol. acetic acid), centrifuged for 5 minutes at 400 g and the pellet is re-suspended in glacial acetic acid solution of 45%. The suspension is then centrifuged. The cells re-suspended in some drops of supernatant are spread on slides.

Remarks

This method involves two critical steps: fixation and air drying. In general, the results are easier in species having small size gonads. In species having big size gonads, as in Bos taurus, it is more difficult to obtain preparations during short lasting stages (as diacinese or metaphase II). Two alternatives are proposed in this case: either the dilaceration of big fragments of gonads in a large volume (1 or 2 litres) of hypotonic solution, or the dilaceration one by one under binocular microscope of several seminiferous tubules (Popescu, 1971).

Certain methods of somatic chromosome banding, e.g. T banding or heterochromatin CBG banding can give good results on meiotic chromosomes (Figure IV.2). QFQ banding method is less reproducible but allows the analysis of heterochromatic regions.

IV.1.2
Technique of synaptonemal complex

Y. RUMPLER and O. GABRIEL-ROBEZ

A particular structure called synaptonemal complex begins to be formed during zygotene stage of the meiotic prophase, just as the homologous chro-

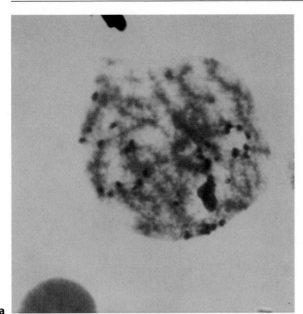

Figure IV.1. Spreading of meiotic chromosomes of male mouse (photographs P. Popescu).
(a) zygotene stage
(b) pachytene stage.

a

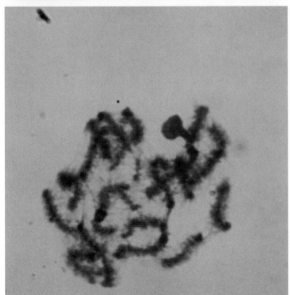

b

mosomes realise the pairing (the meiotic synapsis). This structure is maintained during the whole pachytene stage and disappears at early zygotene when chiasmata is produced.

Two meiotic developments play an important role in the normal spermatogenesis activity: total inactivation of the X chromosome and the synapsis or total pairing of homologous chromosomes.

Figure IV.1. *Continued.*
(c) diplotene stage
(d) diacinese stage.

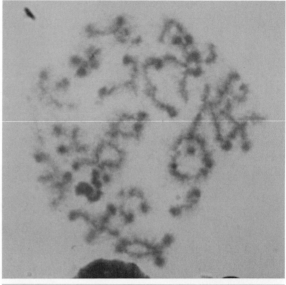

c

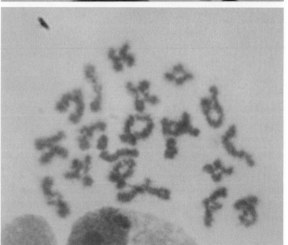

d

Principle

The analysis of synaptonemal complex using electron microscopy is a reliable and useful method to study the different stages of meiotic chromosome pairing and detect the troubles which can be produced during meiosis.

The study of synaptonemal complex is performed on a testicular fragment removed by biopsy under anaesthesia. Germinal cells are extracted from this biopsy and after a hypotonic treatment, the cell suspension is spread on a slide coated by a plastic film and stained using silver nitrate. The spermatocyte I nuclei spreads are first selected with the light microscope for their best-

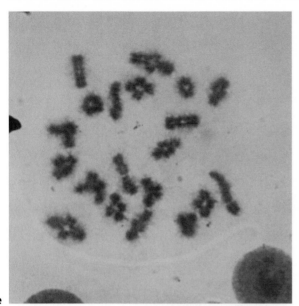

Figure IV.1. *Continued.*
(e) advanced diacinese
stage (f) haploid
metaphase II stage.

e

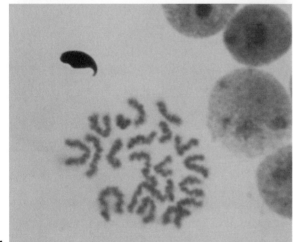

f

spread appearances. They are then transferred, with their plastic slide, onto a grid and analysed using electron microscope.

The microspread method described here is based on the protocols of Counce and Meyer (1973) and Moses (1977).

Protocol
All the glassware used during this handling must be clean, dry and sterile.

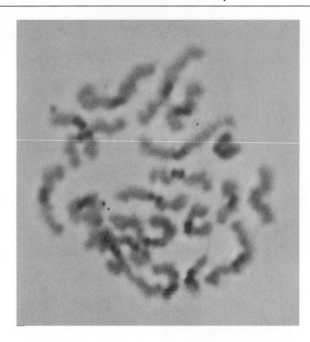

Figure IV.2. Spreading of meiotic chromosomes in mouse metaphase II after CBG banding. Autosomes centromeres and Y chromosome are stained (phase contrast photographs P. Popescu).

■ **Testicular biopsy.** A testicular fragment is removed from an animal under anaesthesia and placed in Ham F10 or Ham F12 medium (appendix M1) containing 10% or 20% bovine foetal serum BFS pre-warmed at 37 °C.

■ **Preparation of cell suspension.** The testicular fragment is placed on a parafilm sheet and blocked with a forceps. The fragment is then finely dilacerated using a scalpel, in Ham F10 or Ham F12 medium (appendix M1) without serum, to obtain a maximum of cell suspension. The suspension is dropped using a Pasteur pipette into a Petri dish. The fragments will pour off for about 20 minutes as the biggest fragments deposit at the bottom. The supernatant (without clots or aggregates) is transferred in a conic tube and centrifuged at 400 g for 5 minutes. The centrifugation supernatant is discarded and the pellet is re-suspended in Ham F10 or Ham F12 medium. The pellet is re-suspended in three-times its volume of fresh medium after a second centrifugation.

■ **Preparation of slides.** The slides are carefully washed (appendix M6). The slides are at first coated by a plastic film immersing them in a solution of chloroform containing 0.5% plastic optilux (appendix M8).

■ **Hypotonic treatment and cell spread.** Three salt-cellars are filled with a hypotonic solution containing agarose 0.2 M (3.4 g in 50 ml of distilled water) prepared and filtered on 0.22 μm Millipore filter, for a maximum of two hours before use. As the cell suspension is variable, three tests of hypotonic treatment must be performed in parallel, placing in each salt-cellar different volumes of cell suspension: 1, 2 or 3 drops are carefully dropped on the surface of hypotonic solution bubble. The cells are then allowed to spread for 30 seconds (the dispersion of the rosée suspension is well localised). The cell suspension is then spread on a slide coated by a plastic film, touching the surface of the bubble with the slide and avoiding touching the edge of the salt-cellar. Pressing must be applied on the centre of the bubble and the slide must be immediately turned up to preserve a small quantity of suspension. The slide is then placed horizontally on a bend-glass support in a humid Petri dish for 5 minutes.

■ **Fixation.** The slide is incubated for 5 minutes in a borel glass containing fixative solution (appendix M9) added with SDS at a final concentration of about 0.03% [20 to 30 drops of 1.5% SDS (weight/vol.) in 50 ml of pure fixative]. The slide is then drained for 5 minutes in a second borel glass containing pure fixative, drained again and finally incubated for 30 seconds in 0.4% Photoflo (appendix M10). The slide is vertically dried, protected against dust by a filter paper.

■ **Observation.** The slide is observed under a contrast phase microscope using a magnification ×10 (avoid the immersion oil on the slide) to observe the cells and the synaptonemal complexes and to control also if the hypotonic treatment was suitable (well spread cells, well situated complexes and very clear cytoplasm).

■ **Staining.** Three drops of 50% $AgNO_3$ solution in distilled water (appendix S13) are placed on the slide. The slide is covered with a cover slip and incubated for 20 minutes at 50–60 °C in dry chamber. The temperature of the incubator should not be over 60 °C, otherwise the solution will crystallize and it will be difficult to remove the plastic film. After this first staining, the slide is dipped in a borel glass filled with distilled water at 4 °C in order to remove the coverslip, then in a second borel glass stirring the slide. After airdrying for at least one hour, three drops of silver nitrate (appendix S14) and three drops of 3% formamid (appendix S15) are placed on the slide. The slide is recovered by a coverslip, placed in a dry chamber and the staining evolution is observed until yellowish and brownish zones appear. At this moment, the staining is blocked by rinsing in water and the slide is dried.

■ **Detection of synaptonemal complexes.** The preparations are observed under phase contrast microscope to detect the regions of the slide containing numerous and well-spread cells. The cells should be well-contrasted and should have a reduced background. The synaptonemal complexes are brownish and the centromeres are dark. Presence of clots, aggregates or spermatozoa tile should be avoided in the microscope field. Slide regions are marked by an indelible (without xylol) fine point marking pen. The microscope objective is slowed round out of the microscope field, the voltage is increased and two points of about 3 mm of distance (grid diameter) are marked. These points enclose the selected pachytene nucleus.

■ **Plastic film floating off.** Floating off must be performed as soon as possible after marking, to avoid adhesion. Cut the plastic film using a razor blade and a clean slide standing back on the preparation slide then mark a stroke, pressing especially on the corners. The slide with the preparation on top is slowly immersed in a glass jar filled with bi distilled water, the plastic film is floated off, avoiding the folding.

■ **Transfer on the grids for observation under electron microscopy.** Cooper-Rhodium mesh 75 are used, handling them always with pliers (clips). Grids are applied on the regions previously detected of the plastic film floating on water, silvered side on the cells. The preparation (the film together with the grids) is carefully fished out of the water with a piece of parafilm, larger than the plastic film. Then, the parafilm is gently pressed to adhere on the plastic film. After 10 minutes the film is hold on the parafilm with a needle, the whole is slowly pulled with a pair of tweezers and flatly placed in a Petri dish. The Petri dish is placed for 2 hours in an incubator at 37 °C. The grids are then taken one by one making small holes with a needle, around each of them, and then flatly placed on clean filter paper, in a glass Petri dish.

■ **Grids observation in optic (light) microscopy.** Grids with the plastic film on the top are placed on a very clean slide to permit the examination under phase contrast optic microscope with magnification ×10. Only the grids containing well-spread and contrasted cells are selected. They can be stored for a maximum of two months in a grid-box.

■ **Observation under electron microscopy.** Spreads are observed under 60kv electron microscope to examine and photograph the well-spread cells having well contrasted and visible centromers and sex vesicle. In case of abnormalities, trivalents and quadrivalents, synapse failures and association with the sex vesicle must be observed.

Film sensitivity must be adjusted before taking several photographs to avoid light variation and problems of connection.

Remarks
This technique of synaptonemal complex analysis allows to study finely the different stages of the chromosome pairing and prophase chronology of the first meiotic division but not those of the other stages of meiosis (Figure IV.3). In (individuals) carriers of chromosome rearrangement, this method permit to analyse the pairing modalities of abnormal chromosomes (Figure IV.4) and to understand the consequences on the spermatogenesis and fertility.

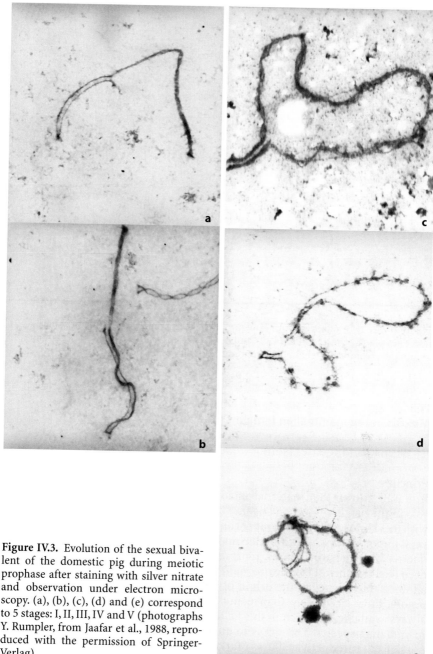

Figure IV.3. Evolution of the sexual bivalent of the domestic pig during meiotic prophase after staining with silver nitrate and observation under electron microscopy. (a), (b), (c), (d) and (e) correspond to 5 stages: I, II, III, IV and V (photographs Y. Rumpler, from Jaafar et al., 1988, reproduced with the permission of Springer-Verlag).

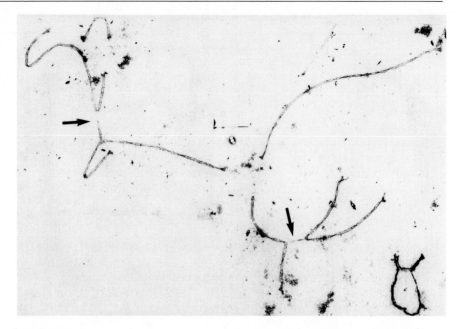

Figure IV.4. Two quadrivalents showing heterospecific pairing (heterosynapsis) of domestic pig carrying both a t(4;14) and a t(3;7) reciprocal translocation. (photographs Y. Rumpler, from Jaafar et al., 1989, reproduced with the permission of Karger AG).

IV.2
Meiosis in the mammalian female

F.J. ECTORS and L. KOULISCHER

Principle

It is well known that maternal meiotic non-disjunction plays a prominent role in the origin of human trisomy 21. Age seems the only factor associated with this abnormality. However chromosomal studies of oocytes are certainly very important to allow a better understanding of the origin of the reproductive disorders or failures, represented either by birth of abnormal children (e.g. trisomy 21), or by pregnancy wastage. Moreover, molecular biology allows today to specify the origin of a chromosomal aneuploidy found in a zygote, more accurately than using cytogenetical markers, which are not always available. Research in this field has been extended both in human and mammalian species. The aim is to understand and explain a significant proportion of reproductive failures (Tarkowski, 1966; Almeida and Bolton, 1993). Studies of female meiosis are much more difficult than those of male due to the anatomic position of the ovaries and the fact that most of the stages of the first meiotic division are produced during the foetal phase. Resuming of the stages of female meiosis occurs at puberty and during adult age, with the end of the first meiotic division at the ovulation, and the end of the second meiotic division after the fertilisation. Consequently, the study of the first stages of the female meiosis requires foetal ovaries.

The protocol of female meiosis study in the adult cow is described here (Ectors et al., 1995).

Protocol

■ **Ovaries collection.** Ovaries are collected from slaughter houses, independently of the oestral (oestrus) cycle and hormonal condition of the adult cow. One hour after slaughter they are dissected and placed at 30 °C in a saline solution (NaCl 0.9% weight/vol.) added with 50 mg/litre kanamycin. The ovaries are brought to the laboratory within two hours, washed 3 or 4 times in a saline solution at 30 °C, then stored in a large neck Thermos flask containing the same medium, before the follicles are collected.

■ **Collection of Cumulus-Oocyte Complexes (COC).** Collection is performed by aspiration with a liquid jet vacuum pump connected to a 5 litre-Erlenmeyer, a mercury manometer, a 50 ml-conic tube and a needle of 19GX11/2 TWB. The depression used is of about 10 to 20 torrs. This depression is achieved through a controlled air leak of the Erlenmeyer. All follicles from about 3 to 12 mm diameter are collected in a conical tube placed between the needle and the liquid jet vacuum pump. Once all the ovaries are punctured, the content of the tube(s) is allowed to sediment for 10 min. The supernatant (follicular fluid) is removed, forming a pellet at the bottom of the tube, the cumulus-oocyte complexes or COC are isolated.

■ **Sorting and washing of cumulus-oocyte complexes.** The pellets are collected and re-suspended in 10 ml of TCM199-HEPES-BSA (appendix CM3). The washing of the cumulus-oocyte complexes (COC) is performed to remove meiotic inhibition from the follicular fluid. The time of this washing is exactly noted, as it is considered the start of the in vitro meiosis. This point is the reference for the subsequent processing of these oocytes. The content of the washing tube is placed in a grid-bottom Petri dish (Falcon 1034) for the observation of the COC under stereomicroscope. Only cumulus oocyte complexes with 3 to 5 coatings of intact and compact cumuli and coronae radiatae, and evenly granulated ooplasm without degenerative changes are selected for maturation (Figure IV.5a). The COC are then washed 3 times in TCM199-HEPES-BSA.

■ **In Vitro Maturation (IVM).** The maturation is performed in 4-holes culture dishes (Nunc 176740) each containing 1 ml of TCM199-FCS-LH-FSH (appendix CM4). 20 to 50 COCs are transferred to each hole, then the culture dish is immediately incubated at 39 °C, in humidified air atmosphere with 5% CO_2. The maturation time lasts 25 more or less 0.5 hours.

■ **Oocyte stripping out from COC.** After IVM the COC are subjected to an enzyme solution followed by a mechanic treatment, to remove the oocytes of

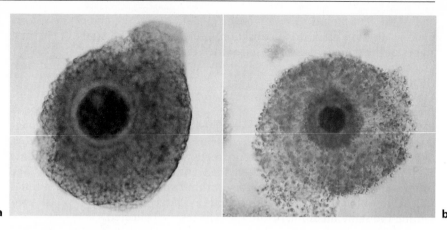

a   b

Figure IV.5. (a) Cumulus oocyte complex selected for in vitro maturation following morphological criteria: an intact unexpended cumulus with min. 3 to 5 layers and an oocyte cytoplasm without degeneration. The cytoplasm (ooplasm) of cattle oocyte is rich in lipids. Therefore, the prophase I nucleus or germinal vesicle, appears like a light spot (magnification × 40) (photographs F. Ectors and L. Koulischer).
(b) Complex "mature" oocyte-cumulus obtained after 25 hours of in vitro maturation and showing very expanded cumulus cells. The corona radiata, more unexpended, is visible, whereas the germinal vesicle has disappeared (magnification × 20) (photographs F. Ectors and L. Koulischer).

the cumulus cells. A solution of 1 mg/ml hyaluronidase (appendix M11) for 10 minutes at 39 °C allows removal of most cumulus cells. Corona radiata cells, less reactive to the enzymatic treatment, are removed by Vortex. The COC, partially denuded are then placed in a tube of 15 ml containing about 3 ml of buffer solution PBS⁻ (appendix B1). The vial is then strongly shaken for 3 to 5 minutes in a Vortex (position 5). One ml of foetal calf serum is added in the tube, to prevent the naked oocytes from sticking to the tube wall. The whole content of the tube is then transferred into a Petri dish to collect the oocytes. These are washed and placed in another Petri dish containing TCM199-HEPES-BSA medium (appendix CM3).

■ **Oocytes fixation and staining (modified method of Tarkowski, 1966).** The oocytes which present a correct maturation (expulsion of the first polar body is the morphologic reference of a correct maturation or at least a maturation beyond the first meiotic division) are selected (Figure IV.5b).

Before the fixation and anticipating the swelling of cytoplasm to obtain the best possible chromosome spreadings, the oocytes are placed at room temperature and transferred into a hypotonic solution of 0.85% KCl (weight/vol., appendix M12). Cells are then placed on a standard slide, previously washed and siliconed, with a minimum of hypotonic solution. The fixative (3 vol. methanol – 1 vol. acetic acid) is added drop by drop, so it is diluted in the hypotonic solution and fixation is progressive. The oocytes, placed on slides, are finally contained in a small drop of fixative. Slides are then allowed to dry at room temperature, before staining with acetic orcein (according to paragraph I.5.2) and covered with a cover slip glued with Histomount. Slides are then examined under immersion microscope (magnification ×1000).

■ **Karyotype analysis of mature oocytes.** The metaphasic spreads are analysed and metaphases I or II are identified (Figure IV.6 and IV.7). Even if all oocytes do not allow thorough counting of chromosomes, whole metaphases can almost always be determined as metaphase of the first or second meiotic division.

IV.3
Study of the spermatozoid by interspecific in vitro fertilisation (insemination)

P. POPESCU

Principle
The analysis of the spermatozon chromosomes must be possible, with the view to have a thorough knowledge of the part taken by the gamet in chromosomal abnormalities, particularly in aneuploidy. The main difficulty in this analysis comes from the fact that the spermatozon chromosomes become visible only after the fertilisation of an oocyte by a spermatozon, at the formation of a male and female pronuclei, before the first division. The mature oocyte contains factors which allow the reactivation of the spermatozon nucleus and the decondensation of its chromatine.

Yanagimachi et al. (1976) succeeded the first heterospecific fertillisation of a hamster oocyte by a human spermatozon with the reactivation of its nucleus, and Rudak et al. (1978) obtained the first chromosomes preparations of a human spermatozon after the insemination of a depellucidated hamster oocyte.

Protocol
In spite of a series of successive modifications of the original protocol of Rudak, the technique is long, taking several days. The protocol described here is mostly based on Séle et al. (1985).

■ **Preparation of spermatozoa.** A semen sample is collected and placed in an incubator of 5% CO_2, at 37 °C for 20 to 30 minutes, to reduce it to the liquid state. Sperm quality is analysed: count of spermatozoa, motility and forward progression. The penetration of spermatozoa into the ova require a capacitation stage (physiological change) of the spermatozoids. This change is performed using the incubation in the appropriate medium, either in a cold or in a warm state.

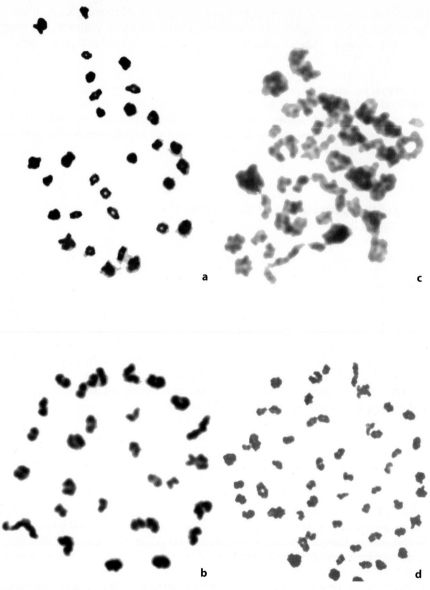

Figure IV.6. Different stages of meiosis in cattle: (a) normal metaphase I; (b) haploid normal metaphase II, (c) metaphase showing many unpaired chromosomes and (d) diploid metaphase II (magnification × 100) (from Ectors et al., 1995, reproduced with the permission of Elsevier Science Inc.).

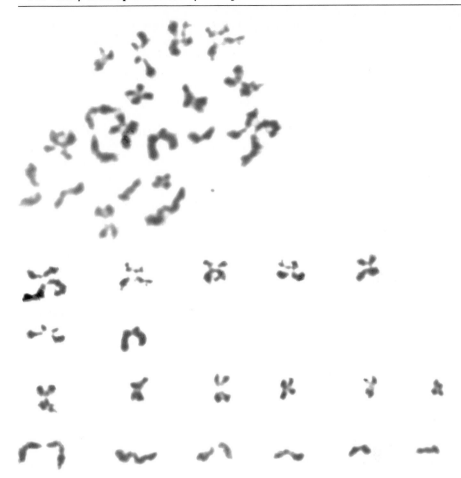

Figure IV.7. Sow chromosome spread in haploid metaphase II and corresponding karyotype (photographs P. Popescu, reproduced with the permission of Expansion Scientifique française).

Protocol of capacitation in the cold state

After the sperm is reduced to a liquid state, 200 µl of sperm suspension are added to 800 µl of TYB medium (appendix CM 5) and incubated for 24 to 72 hours at 4 °C and then placed at 37 °C for 1 hour. The suspension is centrifuged at 600 g for 10 minutes and the pellet re-suspended in 3 ml of BWW working solution (appendix CM6). This operation is twice repeated and finally the pellet is re-suspended in BWW solution at a concentration of 2.10^5 sperm/ml and stored at 37 °C for the insemination.

Protocol of capacitation in the warm state

After the sperm is reduced to a liquid state, 1 ml of the semen is dissolved in 9 ml of working solution BWW (appendix CM6) and centrifuged at 600 g for

10 minutes at room temperature. The pellet is re-suspended in 10 ml of working solution BWW) a second and a third centrifugation are performed under the same conditions. The final pellet is re-suspended at a concentration of 10^6 sperm/ml of BWW medium. The capacitation is performed by the incubation of the sperm suspension in an incubator of 5% CO_2 at 37 °C for 4 to 5 hours before contact with the oocytes.

■ **Oocytes preparation.** Three to four mature female golden hamster (*Mesocricetus auratus* 2n = 44 chromosomes) bred in normal conditions of temperature and lighting are used for each experiment. They are superovulated by one intraperitoneal injection of 50 U.I. PMSG (Pregnant Mare Serum Gonadotropin) the day 1 of the cycle and 50 U.I. of HCG (Human Chorionic Gonadotropin) the day 3 of the cycle. 17 hours after the injection of HCG, the females are killed and their oviducts are collected and suspended in a working medium BWW (appendix CM6). Petri dishes with 30 mm in diameter some of them containing, BWW medium containing 0.1% hyaluronidase solution (appendix M11) and the others 0.1% trypsin (appendix S5) in BWW, are placed on a hot-plate at 30 °C. Each oviduct is dilacerated and transferred to a dish containing hyaluronidase, under binocular lens, using micropipette (Pasteur pipette flame drawn). After 1 to 15 minutes the oocytes become free-cumulus, then they are rinsed in a working solution BWW and placed in a dish containing trypsin to remove the zona pellucida. Oocytes are then washed twice in BWW medium. Oocytes (eggs) are suitable for insemination.

■ **Insemination and oocytes culture.** Drops (200 µl) of a suspension containing motile capacitated spermatozoa are placed in Petri dishes and covered by paraffin oil. 10 to 15 depellucidated oocytes are placed in each drop of sperm suspension and the Petri dishes are incubated in CO_2 incubator for 3 to 4 hours. After the penetration of the spermatozoa nucleus in the oocyte, the male and female pronuclei became visible. The chromatin of the spermatic nucleus is decondensed, the nucleus increases and the membrane disappears. The chromosomes of the oocyte and of the spermatic nucleus are visible without their nuclear envelope which is reformed at a further stage.

After insemination the oocytes are washed twice in the working medium BWW (appendix CM6) with the object of eliminating the spermatozoid suspension, then they are placed in a medium in order to preserve their survival, that is 10 to 15 drops by drop of 100 µl of HAM F10 medium (appendix CM1) containing 15% FCS for 14 hours in a CO_2 incubator. Each drop is recovered with paraffin oil. During the last 6 to 8 hours they are transferred in drops of medium containing 0.2 µg/ml of colchicine (appendix V2).

■ **Hypotonic treatment and fixation.** The hypotonic treatment is performed in a solution of Na_3citrate. $2H_2O$ at 1% (weight/vol.) or KCl at 0.7% (weight/vol.) placed in Petri dishes. Groups of 1 to 10 oocytes are placed for observation under stereomicroscope in hypotonic solution and incubated for 6 to

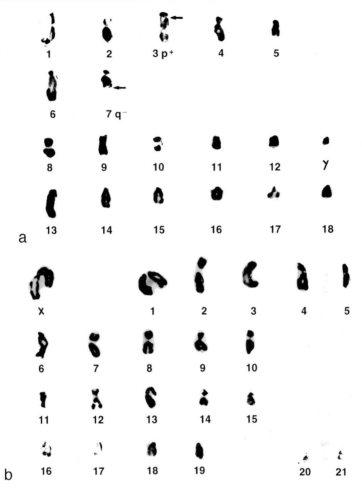

Figure IV.8. (a) Karyotype spermatozoa of a boar carrier of a reciprocal translocation t(3;7).
(b) Karyotype of a depelucidated hamster oocyte after penetration by a boar spermatozoa (photographs P. Popescu, from Benkhalifa et al., 1992, reproduced with the permission of Expansion Française Scientifique).

7 minutes at room temperature (20 °C). They are then transferred in a minimum of hypotonic solution on a clean slide, under stereomicroscope too. A drop of fixative (3 vol. ethanol – 1 vol. acetic acid) is dropped with a pipette on each oocyte group. In this manner the oocytes are dispersed. Each oocyte becomes visible to the naked eye or under stereomicroscope when the fixative begins to evaporate. The second drop of fixative is added, then a third and a fourth. The position of the individualised oocytes is marked on the slide encircling them with a circle using a diamond point or a marker on the back of the slide. 5 to 10 oocytes can be spread on the slide. After drying,

the chromosomes spreads are observed under phase contrast microscope. After two days of maturation a GTG or RHG banding may be performed according to usual protocol.

Remarks

The quality of the technique was improved by the modifications provided to the original method of Rudak et al. (1978). It was impossible to reproduce this method. Good quality chromosome spreads preparation (Figures IV.8a, 8b) depends on several factors: spermatogramme values (number and mobility of spermatozoa), penetration rate of about 30%, the conditions of the culture and control of the cytogenetical techniques.

V
The Lampbrush Chromosomes of Amphibians

F. Simon, N. Vichniakova, C. Pyne and J.C. Lacroix

V.1
Introduction: The basis of lampbrush chromosome mapping

The lampbrush chromosomes correspond to a special organization taken by the chromosomes in the female germ line during oogenesis in several vertebrates and invertebrates, when the oocyte is in the growth phase and its nucleus, the germinal vesicle, is arrested in the prophase of first meiotic division. Nevertheless, these chromosomes are not susceptible to analysis in all the groups and have been studied mostly in amphibians and birds. For the amphibians the lampbrush chromosomes have been mapped essentially in urodeles and a few anourans (Callan, 1986). In the present work we have selected an urodele, the Pleurodele, as a model. This animal reproduces well in captivity and consequently it is often used in embryological studies. The lampbrush chromosomes of Pleurodeles have been used for cytogenetic analysis, especially for the detection of chromosomal alterations. They have also been used for identification of natural populations and isolation of lineages having specific markers on the lampbrush chromosomes (Gallien et al., 1965; Jaylet, 1967 and 1972; Labrousse, 1966, 1970 and 1971; Lacroix, 1968a and b; Lacroix and Loones, 1971 and 1974). The basic method for the preparation of lampbrush chromosomes is the same for all amphibian species, but specific adaptations may be necessary for each species (see Callan, 1986, and Gall et al., 1991).

V.1.1
Organisation of the lampbrush chromosomes

Amphibian oocyte grows into a giant cell (1 to 2 mm in diameter), and all its nuclear and cytoplasmic components are implicated in this gigantism. The large size of the germinal vesicle of the amphibian oocyte (100–200 mm in diameter) has allowed the development of satisfactory preparation techniques for the isolation of the lampbrush chromosomes. These chromosomes are rather fragile, because of their important length and the thinness of their axis. For this reason, manual extraction has to be performed. It has been shown that transcription persists in *in vitro* after the isolation of germinal vesicles (Schultz et al., 1981). The growth phase of oocytes in amphibians may take several months, and during this period the chromosomes remain arrested in the diplotene stage, the so-called "lampbrush" stage. Lampbrush chromosomes can be isolated at any time, from one end of the year to another, as oogenesis is not synchronous. At this stage the homologous chromosomes are paired in bivalents, each chromosome being formed of two juxtaposed chromatids having a lampbrush-like organisation. These chromosomes have an important transcriptional activity, which concerns about 5% of the length of the nucleofilament forming each chromatid. The inactive regions of the two chromatids are apposed together and organised in a single longitudinal structure forming the axis of the chromosomes; this axis has a chromomeric structure. In contrast, the transcribed regions spread on each side of this axis in paired lateral loops. The transcription takes place at the level of these loops. Each homologous is formed by a succession of paired symmetrical loops, emerging throughout the length of the chromomeric axis (Figures V.1a, V.5c, V.9a and V.10). The loops are therefore the functional organisational units of the lampbrush chromosomes. In urodeles the quantity of repetitive DNA in the genome is very high, which might explain the particularly large size of the lampbrush chromosomes in this group. The axis of the larger sized homologous is up to 1500 mm long, and of the lateral loops, up to 300 mm (Figure V.4a, b and c).

These loops can be observed under the phase contrast microscope and counted; there are about a few hundred of them in larger sized homologous, and about 3 to 4000 in the whole haploid genome. All the loops have the same basic organisation: an axis formed by a nucleofilament carrying transcription units. It is known that each transcription unit is constituted by a

---

Figure V.1. Schematic drawings of homologous segments or loops of the lampbrush chromosomes (Figures J.-C. Lacroix).

(a) Median region of bivalent XI of *Pleurodeles waltl* (ibero-lusitanian). Note chromomeric differentiation of the axis and the presence of different types of loops or lateral structures; A – A loops; B – B or globular loops; C – C loops; S – sphere or C snurposomes.

(b) and (c) Two morphological aspects of an A loop, (b) without or (c) after labelling with an antibody which reacts with its matrix and alters its morphology ("immunomorphological labelling").

(d) A typical globular loop observed on the bivalent VI in a genetic male (ZZ) transformed into a female of *P. waltl*. The inversion of sexual phenotype "amplifies" the development of matrices of certain loops. Compared with the state of the same structure in a normal female carrier of a heterozygotic inversion (Figures V.9 and V.10) and with the B loops of bivalent XI (Figure V.1a).

(e) and (f) C loops. One of the loops (e) was localised on the bivalent II of *P. poireti* involved in a reciprocal translocation with a chromosome XI of *P. waltl*, in a female hybrid offspring. The hybridisation also amplifies certain matrices. The other loop (f) represents a "normal loop" which has taken a C loop-like appearance due to an amplification induced by the chimeric nature of the carrier. This loop has actually been observed on the differential segment of the sex chromosome W of a female heteroplastic chimera produced by grafting a head of *P. poireti* onto the body of *P. waltl*. This matrix morphology is rarely observed under normal conditions, but is regularly observed after staining with antibodies which react with its matrix.

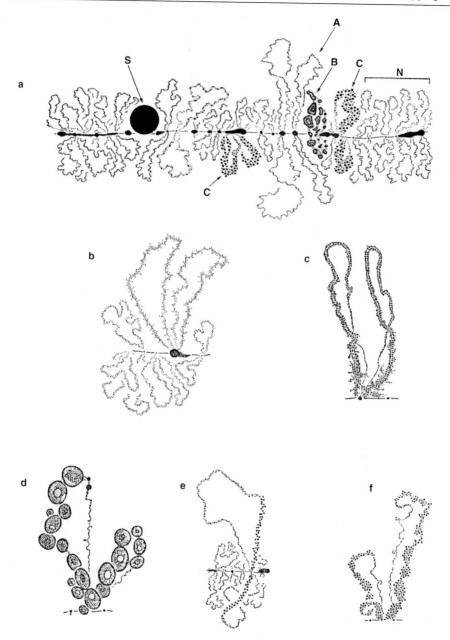

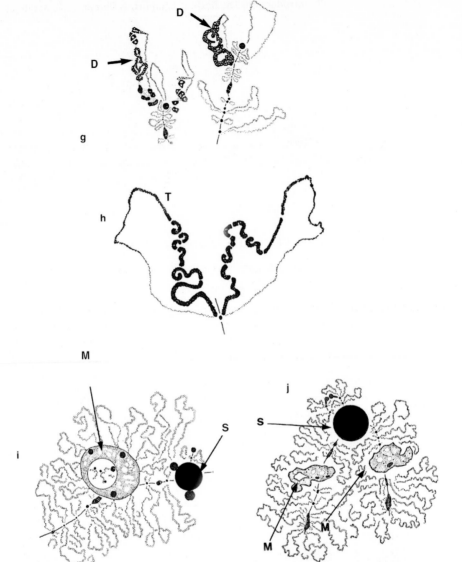

Figure V.2. Schematic drawings of loops and other structures of lampbrush chromosomes (Figures J.-C. Lacroix).

(g) Right end of bivalent II of *P. poireti* showing the variability of D loops in the same bivalent.

(h) Tubular loops (T) localised near the left end of bivalent I of *P. waltl.*

(i) and (j) Left end of bivalent IV, which is the sexual bivalent, in a male transformed into a physiological female (i), and in a normal female (j). Note in (i) the presence of two B snurposomes on the surface of the sphere, which is itself a C snurposome. The M structure is heterozygotic, and only the homologous carrying it is represented. In this case, the M, considered as a heterochromatic structure, is condensed and leads to a reduction of about 70 mm in the length of the chromosome axis carrying it. In (j), the M structures are in a homozygous state.

sequence of nucleoprotein fibrils of increasing lengths, formed by elongating nascent RNA associated with a variety of proteins (Miller and Betty, 1969; Gall, 1991). In situ, the RNP fibrils of a transcription unit are not spread out but are organised in a network around the nucleofilament; this network is known as "loop matrix". The polarity of the matrix reflects the polarity of the transcription unit, i.e., the direction of the progression of RNA polymerase complex along the DNA template, and is observed under the light microscopy (Figure V.4 and V.5b). A chromosomal loop usually contains one transcription unit. However, some loops contain more than one such unit; in this case some of these units may have opposite polarities (Angelier and Lacroix, 1975).

V.1.2
Morphological mapping of the lampbrush chromosomes

V.1.2.1
Morphological variations of genetic origin

The lampbrush chromosomes present a longitudinal differentiation along their axis as well as at the level of the loops. The chromosome axis presents a very important linear differentiation in so far as the successive chromomeres may be different in size and in their affinity for stains, depending on their relative content of DNA and proteins. Certains chromomeres can be easily distinguished from the others and used as chromosome markers. The mapping of these linear differentiations allows a precise observation of chromosomal rearrangements (Figures V.1a, V.9a and V.10). Most of the loops have similar size and morphology; they are known as "normal loops". Certain loops are, however different in size and particularly in the morphology of their matrices; these are commonly used as markers. The morphological characteristics of the loop matrices depend on the specificity of their DNA and consequently the nature of their RNA content, the nature and variety of the proteins associated with their RNA; and the type of release of the ribonucleoproteins (RNPs) from the loops. The morphology of the loops is moreover subject to variations of three origins: genetical, physiological and developmental stage (Callan, 1963), i.e., the morphological specificity of the loops reflects the specificity of their DNA. Actually it has been proved that the characteristics of the loops are hereditary and transmitted as simple Mendelian characters when they are carried by autosomes (Callan and Lloyd, 1956 and 1960) and as characters depending on the sex when they are carried by the sex chromosomes (Lacroix, 1970). But this genetic specificity, as observed in the existence of morphological variants occupying the same chromosomal site, can be partially masked by morphological variations of physiological origin. The best identification of genetic variants is done observing them in heterozygous animals: indeed, in each oocyte the two heterozygous loops localised on homologous sites of a bivalent are in the same physiological condition and at the same stage of differentiation of the oogenesis (Figure V.5b).

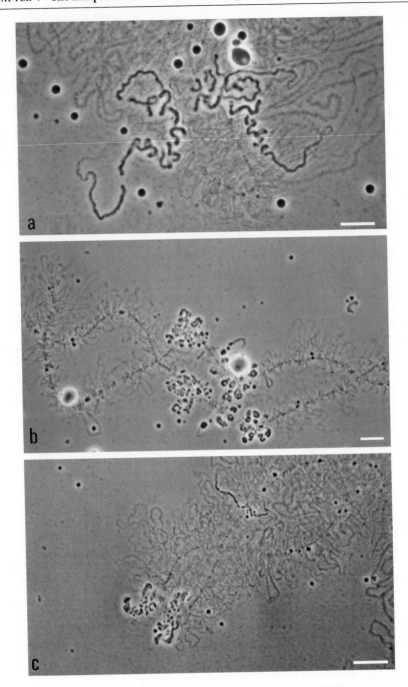

V.1.2.2
Morphological variations of physiological origin

Due to their intensive transcriptional activity, the loops (and consequently the chromosome axis) are very susceptible to factors producing physiological variations. In Pleurodeles there are several factors which produce morphological alterations in the lampbrush chromosomes: irradiation by the γ rays (Loones, 1979), heat shock in vivo (Rodriguez-Martin et al., 1989) or in vitro (Flannery and Hill, 1988; Lacroix et al., 1990), inhibitors of the transcription and biological hormones etc. These alterations are produced according to precise chronologies which have been determined; thus, it is possible to induce the development of certain matrices and to obtain suitable markers. However, morphological variations of the lampbrush chromosomes are also observed after a change in the animal's mode of life (e.g., from the wild state to captivity), interspecific hybridisation, inversion of sexual phenotype, and in heteroplastic chimeras (Lacroix, 1968b). Variations related to the developmental stage of the oocytes are comparable to those of physiological origin. Thus the degree of development of a loop or of a structure like the sphere varies in the course of oogenesis.

In conclusion, the loops, due to their differentiation related to their genetic specificity, form excellent topographic markers for identification of bivalents. They constitute the basis of the mapping of lampbrush chromosome (Callan and Lloyd, 1975). In contrast, the linear differentiation of the chromosome axis is an important tool for comparative analysis of chromosome segments, but it is difficult to realise detailed mapping of the axis of different chromosomes taking into consideration the important physiological variations.

V.1.3
Immunomorphological mapping

Even under the most favourable conditions, the number of chromosomal loops identified by their morphology is less than a hundred on the three or four thousand loops present in the haploid karyotype of an urodele. Consequently, it is necessary to find other methods of detection to develop the mapping of the lampbrush chromosomes, as all the loops, including the normal ones, are potential markers in so far as each of them has a genetic specificity. It is therefore necessary to develop a mapping using labelling by nucleic acid or protein probes, to compensate deficiency of morphological

◀──

Figure V.3. Phase contrast micrographs of different bivalents (micrographs J.-C. Lacroix, bar = 10 μm).
(a) Left end of bivalent I in *P. waltl* (Moroccan) showing tubular loops observed in oocytes cultured for 4 to 6 days. Note the continuity of the matrix of these loops.
(b) Median region of bivalent V in *P. poireti* showing two groups of B loops characteristic of this bivalent. These loops have been amplified by heat shock.
(c) Left end of bivalent II of *P. waltl*, showing the terminal D loops and the subterminal tubular loops.

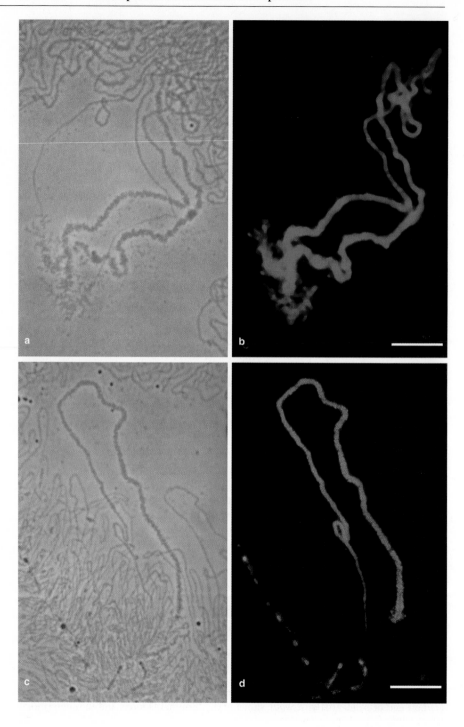

markers. These methods, using in situ hybridization with nucleic acid probes or labelling with antibodies have not been much used until now for the mapping of the lampbrush chromosomes. The basis of a mapping using immunocytochemical labelling is described in this chapter.

Indeed, if the genetic information contained in loops is specific for each unit, some of the proteins associated with the transcription complexes could be specific for certain units. The observations of Scott and Sommerville (1974) showed effectively the existence of proteins specific for certain chromosome units. Studies using monoclonal antibodies raised against proteins of amphibian oocyte nuclei have confirmed that certain monoclonal antibodies label a limited number of chromosome loops, or even only one transcription unit in a loop containing more than one of such units (Figures V.4, V.5a and b) (Lacroix et al., 1985; Roth and Gall, 1987). This labelling is sensitive and reproducible, as it is reliable and efficacious.

The development of antibodies, especially of monoclonal antibodies, raised against the proteins of lampbrush chromosomes is the most rational strategy. We have also observed that antibodies of very different origins, including the natural antibodies of non-immunised animals label the lampbrush chromosomes in a differential and perfectly reproducible manner. Moreover, a loop indistinguishable under phase contrast may be identified after its staining with particular antibodies, not only due to its staining but also because it acquires a morphology different from its initial morphology, i.e., in the absence of the antibodies (or in controls with other non-staining antibodies). Labelling with antibodies is an efficient tool for identification and mapping of certain loops. This procedure is called "immunomorphological mapping" (Figure V.4).

◄───

Figure V.4. A loops with modified morphology by an antibody raised against human P 52 (Figures J.-C. Lacroix, bar = 10 μm).
(a) and (b). A loop localised on bivalent II of *P. waltl.* This loop, observed in phase contrast (a) and in fluorescence (b), is subdivided in 2 comparable sectors, each of them corresponding to a presumed transcription unit. One of them, labelled with the antibody, is characterised by the asymmetric distribution of the ribonucleoprotein particles along its axis and by its "blown up" aspect at the end presumed to be the site of release of the transcription complexes. The other sector, unlabelled, has a very reduced matrix but presents nevertheless the same polarity as the labelled unit.
(c) and (d). A loop localised on the chromosome II of *P. poireti.* This loop appears to be formed of only one transcription unit which covers almost the whole length of its axis. It belongs to the same family of loops as the previous one: it has the same immunological and morphological characteristics.

V.1.4
Maps of the lampbrush chromosomes of Pleurodeles

V.1.4.1
The conditions of maps development

The variations concerning the stages of development or the physiological conditions of the oocyte have an effect upon the morphology of the markers. This variation amplifies or removes them. It is therefore necessary, for the maps development, to operate in optimal conditions under which the morphological markers are the best developed. The experiment is easily accomplished, once the chronology of the markers development is well determined.

The oogenesis in Pleurodeles is divided into six stages (Bonnanfant-Jaïs and Mentré, 1983). Chromosomal structures developed in later stages are less or not developed in earlier stages, and vice versa. The reference chromosome maps presented here have been established from oocytes of stages III to VI, which are more easy to manipulate than the oocytes in earlier stages. The temperature of breeding (20°C) and of oocyte culture (19°C), as well as the procedures of oocyte culture and of chromosomes isolation are also precisely controlled.

Marker structures, which are absent or less developed in normal conditions, but appear well developed in special experimental conditions are also included in these maps. This is the case of certain B loops which are induced by heat shock (Figure V.3b), or tubular loops (T loops) which are visible only in preparations from oocytes having undergone a prolonged culture of 4 to 6 days (Figure V.3a). It is also known that interspecific hybridisation (Figure V.1e, inversion of sexual phenotype (Figures V.1d and V.2i) and production of heteroplastic chimera (Figure V.1f) "amplify" the matrices of all loops and make them more easily identifiable.

V.1.4.2
The intraspecific maps

The two species of the genus Pleurodeles, *P. waltl* and *P. poireti*, are very similar because they can produce fertile interspecific hybrids. Moreover, in

---

Figure V.5. A loop with modified morphology by an antibody raised against human P52 (figures J.-C. Lacroix, bar = 10 μm).
(a) Median region of bivalent V of *P. waltl* (Moroccan). This marker is sufficient to identify this bivalent. Note the chromomeric aspect of the DAPI-stained axis.
(b) Median region of bivalent V, of a heterozygous animal for the same loop; *P. waltl* (iberolusitanian). A dense fluorescent granule is localised just on the right of the insertion point of the labelled loop, at the junction of the two homologous chromosomes.
(c) Median region of bivalent IV of *P. poireti*. This region carries only normal loops. In this micrograph, phase contrast image and the blue fluorescent image showing the DAPI staining are superposed. The two images underline the chromomeric organisation of the chromosome axis. The axial granules poor in DNA appear in black.

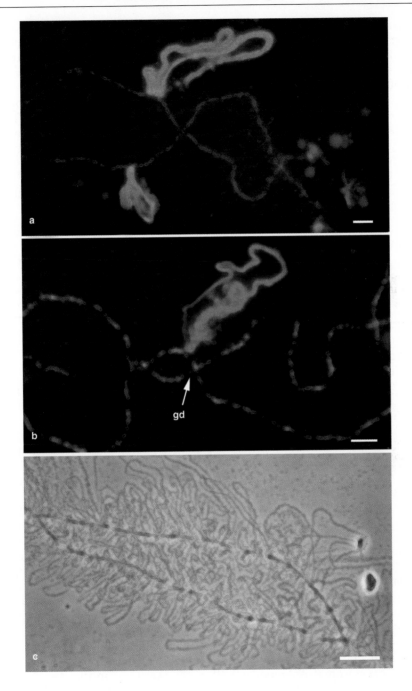

hybrids the homologous chromosomes of the two species form normally paired bivalents in diplotene, which has made it possible to establish homologies between the chromosomes of the two species. Nevertheless, morphological and immunomorphological variants can be observed and allow the identification of populations inside each species.

Actually, taking into consideration the populations bred in laboratories, three maps are enough to characterise the known populations. They correspond respectively to ibero-lusitanien and Moroccan populations of *P. waltl* and Algero-Tunisien population of *P. poireti* (Figure V.12a, b, c). These maps have been reduced to essential markers, to identify each bivalent and to differentiate the population of the three origins. They are based on the first maps of Pleurodeles, which are retaken and simplified. (Lacroix, 1968a).

V.2
Techniques for the preparation of lampbrush chromosomes for optical microscopy

V.2.1
Preparation of oocytes

Protocol
■ **Biopsy of ovaries.** The animals are reared at 20 °C. After anaesthesia with MS 222 (ethyl-m-aminobenzoate) (Sigma A-5040) 1% in tap water (weight/vol.), the female is washed with soap, rinsed with tap water and placed on one of its sides. The other side, where the surgical intervention is to be performed, is wiped with 75% alcohol. A longitudinal incision (1 to 1.5 cm) is made in the abdominal wall, using Dowell scissors, half-way between the ribline and the lateral line. A fragment of an ovary (about 1 ml, containing around 200 to 300 oocytes) is taken out of the wound with a forceps, excised and placed in a saline solution, L1 (Wallaceet et al., 1973) (appendix L1). The rest of the ovary is replaced inside the abdomen; the incision is powdered with sulphonamide and sutured with a silk thread. On waking, the animal is placed in a humid chamber, or replaced in an aquarium. In 2 to 3 weeks the wound is entirely healed. The biopsy can be repeated on the same animal 2 or 3 times a year. The animals possessing interesting markers on their lampbrush chromosomes are reproduced in order to develop desired lines of descendants. The fragment of the ovary can be used directly for the chromosomes isolation, or be treated for the defollicularisation of its oocytes.

■ **Defollicularisation and culture of oocytes.** The defollicularisation facilitates chromosomes isolation, as well as the injections in the oocytes; this operation is necessary in experiments where the follicular cells are liable to interfere. The ovary fragments are rinsed in three successive baths of the saline solution L1 (appendix L1), then the follicles are separated in groups of

3 or 4 with two watch maker's forceps no. 5, under a dissecting stereomicroscope. The follicles are then incubated in a solution of collagenase I, 1 mg/ml (appendix L2), for 4 hours at 25 °C, stirring periodically (4 to 6 times for 4 hours). The follicles are then rinsed twice in the solution L1 (appendix L1) and placed in a solution L3 containing 1 mM EDTA (appendix L3) at room temperature for 15 min., stirring gently and constantly. This treatment removes the follicles and frees the oocytes. From this moment onwards, the oocytes are handled with a Pasteur pipette of 1.5–2 mm in diameter. The defollicularised oocytes are rinsed twice in the solution L1 (appendix L1), strictly sorted, to remove all damaged oocytes showing the least trace of alteration in their pigmentation, and placed in Petri dishes according to their sizes. They are cultured at 19 °C in the solution L1 (appendix L1); gentamycine may be added to this solution at 5 mg/ml. The conservation time of the oocytes in this saline medium varies from a few days to 8 weeks in the best case. The oocytes are used for chromosomes isolation, either directly, 16 to 20 hrs after defollicularisation, or after a heat treatment before the chromosome isolation.

■ **Choice of isolation time.** The chromosomes can be isolated from the oocytes as soon as they are cut off the animal. However, their isolation after oocytes defollicularisation is often preferable, indeed necessary, particularly for experimentation ex vivo. On the other hand, the evolution of the physiological pattern (range) of the chromosomes is chronologically variable, that is, the direction of the evolution is known, while its chronology is variable. Consequently, preparations are made from oocytes from one to six days after defollicularisation. A heat shock may be added: oocytes, maintained at 19 °C for 20 hours after defollicularisation, are placed in an incubator at 37 °C for 2 to 3 hours and then replaced at 19 °C for 1 to 4 days until the isolation. Preparation of chromosomes under these different variants ensures the development of markers necessary for the identification of certain chromosomes.

V.2.2
Chromosomes preparation for morphological studies

Protocol
■ **Extraction of the germinal vesicle.** This is carried out under a dissecting stereomicroscope, in a watch glass containing sterile saline solution L4 (appendix L4), using two watch-maker's forceps no. 5 (Dumont). Two to three oocytes (or follicles) are treated per watch glass. The oocyte membrane is torn using two forceps. The cytoplasm content is freed in a white clot, in which a protrusion is projected, marking the position of the germinal vesicle. The translucent germinal vesicle is freed by squirting gently the saline solution onto it, using a pipette (0.5 to 1 mm in diameter). Then it is washed by several suctions and discardings into the pipette. When the germinal vesicle is quite clean, it is transferred using another clean pipette, to the isolation chamber containing a saline solution L5 (appendix L5).

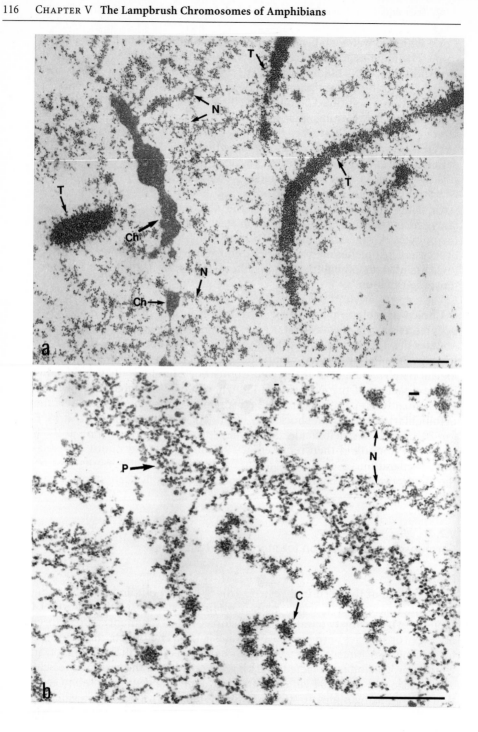

■ **Chromosomes isolation.** This is performed in an isolation chamber consisting of a 6 mm hole perforated glass slide. A coverslip (22 × 32 mm) is sealed onto the slide with paraffin (melting point 42–44 °C). The margins of the coverslip are sealed with transparent nail polish; this coverslip forms the bottom of the chamber. Before use, the chamber is UV sterilised for 20 min to avoid bacterial infection, especially if the preparation is not fixed immediately. Chromosomes isolation is the most delicate step. It is carried out under the stereomicroscope, using a forceps and a tungsten needle. The saline solution L5 (appendix L5) must form a convex meniscus above the opening of the isolation chamber. The germinal vesicle, which lies at the bottom of the chamber, is lifted with the forceps at half way in the chamber, and the envelope is torn with the tungsten needle. This manipulation must be performed carefully, to allow the nuclear clot to spread out at the bottom of the chamber. The nuclear envelope is then removed, or if necessary, placed at the periphery of the chromosomes at the bottom of the chamber. The chamber is placed in a Petri dish on crushed ice for at least 10 min. During this time the nuclear clot disperses and the chromosomes spread out. The state of the chromosomes can be observed at this stage under an inverted phase contrast microscope.

■ **Loops spreading.** The isolation chamber is covered with a coverslip (20 × 20 mm), placed horizontally over the chamber, taking care to avoid the formation of air bubbles. The liquid in excess is gently wiped using a filter paper (Kimwipe), to prevent the coverslip from slipping during centrifugation. The chambers are then placed in the slide-supports of a centrifuge. The preparations are centrifuged at 500 g for 5 min, then the speed is increased progressively at 3000 g, for 25 min, at 6–8 °C. The loops spread out and adhere firmly to the bottom of the chamber. The chromosomes can be observed at this stage under a phase contrast microscope turning over the bottom of the isolation chamber in the upright position.

■ **Fixation.** Different types of fixations can be carried out, according to the further treatment. Only paraformaldehyde fixation routinely used is described, for observations under phase contrast microscopy and for immunocytochemical labelling. The procedure is the same for all kinds of fixation. The chambers are immersed in freshly prepared 4% paraformalde-

←————————————————————————————————

Figure V.6. Electron micrographs of lampbrush chromosomes spread on a coverslip and sectioned by ultrathin sections in parallel direction to the plane of spreading. (Micrographs C. Pyne; bar = 1 μm).
(a) Axial region of a lampbrush chromosome of *P. waltl* (ibero-lusitanian). Note the axis of the chromosomes (Ch), the matrices of normal loops (N) formed by particles of 25–30 nm in diameter, and tubular loops (T) formed by densely packed particles of 45 nm.
(b) Higher magnification image of normal loops (N), granular loops (C) and P loops formed by particles of 45 nm in diameter organised in a matrix less dense than that of the T (tubular) loops.

hyde in Amphibian Ringer solution, ph 7.4–7.5 (appendix B12) and the coverslips removed pushing them gently with forceps for 20 min at 6–8 °C. Preparations are then rinsed in TBS1× (appendix B11), the chambers are covered with coverslips (20 × 20 mm) sealed with nail polish, and examined under phase contrast microscope for morphological analysis. However, the preparations are in general used for immunostaining.

V.2.3
Chromosomes immunolabelling

Protocol
The procedure carried out routinely is described here, taking into consideration that the protocols may be widely modified. All the operations are carried out at 6–8 °C on a refrigerated bench top, as already described. A very delicate step is the wiping of the isolation chamber, to change the medium. Liquid in excess around the hole of the chamber is carefully absorbed, avoiding that the chamber becomes empty, and the chromosomes come into contact with the air. Otherwise, they are irremediably damaged. The liquid is gently absorbed with filter paper (Kimwipes), in such a way that the margins of the chamber are not touched. The chambers must be maintained constantly in a horizontal position. The chambers are immersed and rinsed in three successive baths of 10 min each in TBS1× (appendix B11).

■ **Application of the first antibody.** After fixation the chromosomes are rinsed in three successive baths of TBS. The rinsing solution TBS1X (appendix B11) is replaced by 10% foetal horse serum in TBS1×, for 10 min, to saturate the non-specific sites. Liquid in excess around the chamber is absorbed before the application of the antibody solution. Ascites fluid or polyclonal serum are added at 1/250 (vol./vol.) dilution in TBS; hybridoma supernatant solutions are used undiluted. After an incubation of 30 to 60 min, the preparations are rinsed in three successive baths of TBS1×.

■ **Application of the second antibody.** The second antibody, raised against the first, is either conjugated to a fluorochrome (Texas Red, fluoroscein) or biotinylated and diluted at 1/50 (vol./vol.). It is applied, after wiping, for

───➤

Figure V.7. Electron micrographs of structures associated with the lampbrush chromosomes. [(a), (c) and (d) from Pyne et al., 1989, reproduced with permission of Springer-Verlag; (b) micrograph C. Pyne; bar = 1 μm)].
(a) The M structure, in subterminal position on the left end of bivalent IV of *P. waltl* (ibero-lusitanian). Note the heterogeneous structure of M, with more dense regions.
(b) Sphere (C snurposome) in terminal position on the right end of the bivalent IV. Note the presence of B snurposome (snB) on its surface.
(c) D loop, localised at the left end of bivalent X.
(d) Globular loops (B) localised at the limit of the 1/3 right region of bivalent VII of *P. waltl*.

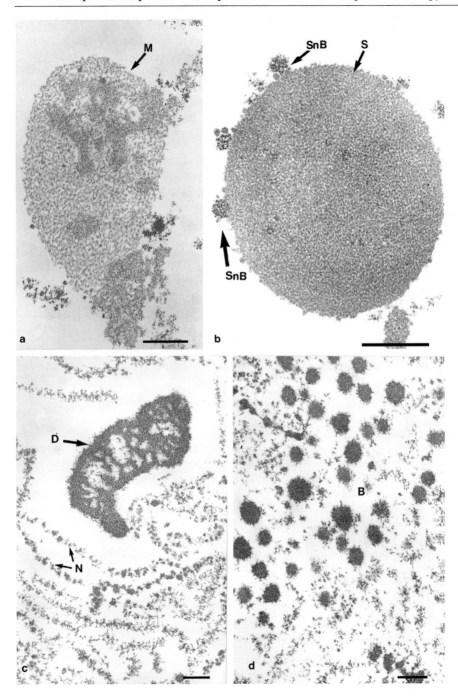

30 min. The preparations are then rinsed. If a biotinylated second antibody is used, the preparations are rinsed and stained with streptavidin-Texas Red at 1/50 (vol./vol.) for 30 min and rinsed again several times.

■ **DAPI counterstaining of the axis.** DAPI (Boheringer, Mannheim, Germany) is used at 1 : 50 or 1 : 100 dilution of a 1 mg/ml stock solution in distilled water (appendix S2). It is added either to second antibody fluorochrome or conjugated to streptavidin-Texas Red solution when a biotinylated second antibody is used.

■ **Mounting and conservation.** After rinsing and wiping around the isolation chamber, the preparation is mounted in a solution of glycerine-TBS, 2 : 1 (vol/vol), pH 7.8. The chamber is covered with a coverslip (20 × 20 mm) and the latter sealed with nail polish. The preparations can be stored in the refrigerator (4–8 °C) for several months.

Texas Red is preferable to fluoroscein as fluorochrome, as it is much less subject to fading, and it combines well with DAPI for a double fluorescence. The staining patterns are perfectly reproducible when the same preparatory conditions are used. The use of ethanol as fixative, saline solutions of different osmolarity, and ultraviolet irradiation of chromosomes are important factors that change the labelling pattern of certain antibodies. It is therefore necessary to use standardised conditions.

V.3
Preparation of lampbrush chromosomes for electron microscopy

The procedure of lampbrush chromosomes preparation for electron microscopy is exactly the same as for optical (light) microscopy up to the loop spreading by centrifugation (paragraph V.2.2). Afterwards, taking into consideration the large sized chromosomes, the selected chromosome or region of the preparation to be sectioned with the ultratome must be very exactly localised in the final resin bloc. For the next steps, different techniques are used according to ultrastructural studies or high resolution immunolocalisation. In electron microscopy, the same precautions for manipulation must be taken as for optical microscopy: i.e., to avoid the chromosomes coming into contact with the air. Note also that the chromosomes are sectioned in a parallel direction to the plane of spreading.

V.3.1
Ultrastructural studies

If the structure to be analysed is localised in phase contrast, the preparations are fixed for observation under electron microscopy, after centrifugation. In contrast, if the structure is localised only by immunolabelling, the fixation of preparation for observation under electron microscopy is carried out after immunolabelling. A first fixation of the immunolabelling must be carried out for the observation under photon microscopy.

Protocol

■ **Chromosome fixation and observation.** Chromosome preparations are fixed after the loops spreading (centrifugation) (paragraph V.2.2.) or DAPI staining of the axis (paragraph II.3.1.4), in a solution of 2% glutaraldehyde in phosphate buffer, 0.1 M, ph 7.4 (appendix B10), for 30 min at room temperature, and rinsed 3× in the same buffer for 15 min. each. The chambers are then covered with a coverslip for observation under phase contrast. A schematic drawing of the preparation is done, and the relative position of the region to be analysed noted. The selected region is then photographed, with the object of reading the electron micrographs.

■ **Embedding.** The coverslip carrying the chromosomes (which forms the bottom of the isolation chamber) is removed with a razor blade. The chromosome preparation is then dehydrated in a series of ethanol baths of increasing concentrations up to 95%. The paraffin adhering to the coverslip is removed with a scalpel under a dissecting streromicroscope handling carefully, taking care not to touch the region containing the chromosomes. The dehydration is completed by three baths of absolute ethanol, 20 min each. The coverslip is then placed in a mixture of absolute ethanol and Epon, 1:1 (vol./vol.), for 2 hours, then in two baths of pure Epon mixture also for 2 hours each. The coverslip is then placed, the chromosomes underneath, over a Beem capsule filled with Epon. The resin is then polymerised at 60 °C for 48–72 hours. (Pyne et al., 1989).

■ **Localisation staining, ultrathin sectioning and contrast staining.** The coverslip is easily removed from the Epon bloc using a razor blade, after immersing the bloc in boiling water for 1 min and then in liquid nitrogen for 30 to 45 sec. The chromosomes, embedded at the surface of the Epon bloc, are stained in a solution of 0.5% sodium borate containing 0.5% (weight/vol.) methylene blue and 0.5% (weight/vol.) Azure II, for 10 min at 60 °C. This staining permits the localisation of the selected chromosome or region under the ultramicrotome. Thin sections are cut with a Diamond knife, and stained with a 1:1 (vol./vol.)) mixture of saturated aqueous solution of uranyl acetate and absolute ethanol for 5 min, to increase the contrast under electron microscope.

V.3.2
High resolution Immunolabelling

Principle

Secondary antibodies conjugated to 5 or 10 nm colloidal gold particles are used for the immunolabelling under electron microscope, at higher

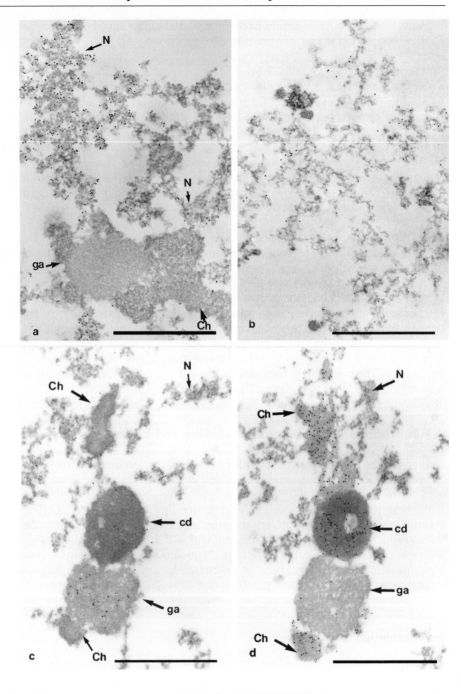

concentrations than those used for the immunofluorescence microscope (photonic). The general principle for the pre-embedding immunolabelling is to perform a simultaneous immunostaining for optical and electron microscopy, and then to post-fix and embed the preparations for electron microscopy (Pyne et al., 1994). For post-embedding immunostaining, the chromosomes are fixed in paraformaldehyde and embedded in LRwhite resin, before the thin sectioning and immunostaining (Pyne et al., 1995).

V.3.2.1
Pre-embedding immunolabelling

Protocol

■ **Application of antibodies.** The protocol is according to paragraph V.2.3 (photonic microscopy), up to DAPI staining (according to par II), except for this modification: 1% (weight/vol.) bovine serum albumin is used as blocking agent instead of foetal horse serum.

The first antibody (ascites fluid or polyclonal serum) is used at 1:50 or 1:100 dilution (vol./vol.) in TBS1× (appendix B11) containing 1% bovine serum albumin, for 30 to 60 min.

The second antibody, biotinylated, is used at 1:50 dilution (vol./vol.) in TBS for 30 min.

The third solution contains a mixture of colloidal gold conjugated Streptavidin at 1/10 or 1/25 dilution (vol./vol.), Streptavidin-Texas Red at 1:50 dilution (vol./vol.) and DAPI at 1:50 dilution (vol./vol.) of a 1 mg/ml stock solution (appendix S2), in TBS; it is also applied for 30 min.

This simultaneous labelling for immunofluorescence and electron microscopy permits an observation of the preparations under epifluores-

◄─────────────────────────────────────

Figure V.8. Colloidal gold labelling of normal loops and axial structures of lampbrush chromosomes [(a) from Pyne et al., 1994; reproduced with permission of Springer-Verlag; (b) micrograph C. Pyne; (c) and (d) from Pyne et al., 1995; reproduced with permission of Elsevier; bar = 1 μm].
(a) and (b) Colloidal gold labelling of the lampbrush chromosomes with pre-embedding procedure. The monoclonal antibody used in (a), A 33/22, labels the normal loops. The loop matrices are labelled; the chromosomes axis (Ch) and the axial granules (ga) show only background labelling. The antibody used in (b), raised against human protein P52, labels the A loops, changing their morphological aspect under light microscopy; the colloidal gold particles are localised on the loop matrix.
(c) and (d) Colloidal gold labelling of axial structures of lampbrush chromosomes, realised after LRwhite embedding of chromosomes. The monoclonal antibody utilised in (c), MPM2 (provided by P.N. Rao, Houston, Texas, U.S.A.), has been raised against mitotic cells. Note the labelling of the axial granule (ga) and its absence on chromosome axis (Ch) and dense chromomere (cd). The monoclonal antibody AC-30-10 utilised in (d), is an anti-DNA which labels the chromosome axis (Ch) and the dense chromomere (cd). In contrast, axial granules (ga) are not labelled.

cence, and the selection of chromosomes or regions to be examined under the electron microscope.

■ **From post-fixation to thin sectioning.** The same protocol as that described for ultrastructural studies (paragraph V.3.1).

V.3.2.2
Post-embedding immunolabelling

Protocol
For flat-embedding of chromosomes in LRwhite, it is absolutely necessary to avoid the resin coming into contact with the air during its polymerisation. For this reason gelatine capsules are sealed with nail polish (or any resin like Araldite), to a bored plastic slide, the margins of the capsule being at a level with the upper surface of the slide. After routine dehydration the preparations are rinsed in two baths of resin. The coverslip is then placed, with the chromosomes underneath, over a capsule filled with resin, taking care to avoid the formation of air bubbles; a second coverslip of larger dimension is placed over it and the edges sealed with nail polish. Polymerisation is carried out at 50 °C for 36 to 48 hrs. The coverslip is removed from the resin bloc by alternative rinsing of the bloc in running hot and cold water, or by placing the coverslip on a flat dry-ice surface after heating the bloc in running hot water. Thin sections of LRwhite embedded chromosomes are picked up on nickel grids, hydrated in TBS1× (appendix B11) for 2 hours, treated with 1% (weight/vol.) TBS1× for 15 min and immunostained. To diminish (decrease) background labelling, which can be high with LRwhite sections and certain antibodies, Tween-20, 0.1%, is added to TBS1× used for rinsing (Pyne et al., 1995).

V.4
Preparation of mitotic chromosomes

Pleurodeles has been the subject of cytogenetic research concerning polyploidy induced by heat treatment and chromosomal rearrangements produced by irradiation (Gallien et al., 1965; Labrousse, 1966, 1971; Lacroix, 1968b; Lacroix and Loones, 1971, 1974; Jaylet, 1967, 1972). The reference karyotype has been established from prometaphase chromosomes from epithelial cells of embryos and larvae (Gallien et al., 1965). It was improved by the induction of secondary paracentromeric constrictions produced by hypothermic treatment (Labrousse, 1970). Karyological analysis has also been carried out on adult tissues: blood, testes and intestine. The procedures used in these studies are basically the same: blocking of mitosis by colchicine, fixation with acetic Carnoy solution and staining with acetic orcein (see paragraph I.5.2) or Giemsa stains (see paragraph I.5.1). Even though the lampbrush chromosomes offer a better resolution for cytogenetic analysis than mitotic chromosomes, the latter are useful in certain cases. Indeed, the mitotic chromosomes are indispensable when the localisation of centromeres is necessary for analyses of chromosomal rearrangements, as markers for centromeres of lampbrush chromosomes are not yet available; the pericentric inversion of chromosome VI provides an example (Figures V.9, V.10). The detection and analysis of polyploidy is also easier with mitotic chromosomes (Figure V.11). We describe here the most easy and rapid method, based on mitosis in embryos or larvae, stained with DAPI or Hoechst 33342 and analysed in epifluorescence.

Protocol

■ **Treatment using colchicine.** Embryos or larvae are placed *in toto* in their breeding medium, to which 0.5% (poids/vol.) colchicine is added, for 5 to 20 hours. If the animal is to be preserved, only the caudal fin behind the anal region is used.

■ **Hypothermic treatment.** To produce secondary constrictions, embryos or larvae are placed in their breeding medium containing colchicine and maintained at 2 °C for 2 to 72 hrs.

■ **Fixation and squashing.** Embryos or larvae are fixed in acetic Carnoy solution (3 : 1 ethanol-glacial acetic acid vol./vol.) for 2 to 4 hours. It is placed on a glass (histological) slide, and epidermal fragments are dissected out from the surface of the embryo or the larval fin using a forceps and a needle. These fragments are then treated with 45% acetic acid for 2 to 3 min, covered with a coverslip and pressed with the palm so as to break the cells and to spread out the chromosomes.

■ **Staining and mounting.** In order to remove the coverslips and to preserve the chromosomes on the slides, the preparations are placed on a flat dry-ice surface (bench) for 10 to 20 min. The coverslip is removed with a razor blade. The preparations are rinsed rapidly in Carnoy solution, to remove cytoplasmic fragments, and then placed in TBS1× (appendix B11). They are stained for 10 min in DAPI (1/50 dilution of 1mg/ml stock solution in TBS1×) (appendix S2). After rinsing with TBS1×, the preparation is mounted in a mixture of glycerine and TBS1× (2 : 1 vol./vol.), pH 7.8, covered with coverslips and the margins sealed with nail polish.

V.5
Analysis of lampbrush chromosomes

The chromosomes are observed under a phase contrast microscope equipped with an epifluorescence device. Appropriate filters are used to observe the fluorescence of the axis stained with DAPI, and that of the loop matrices stained with fluorochrome-conjugated antibodies.

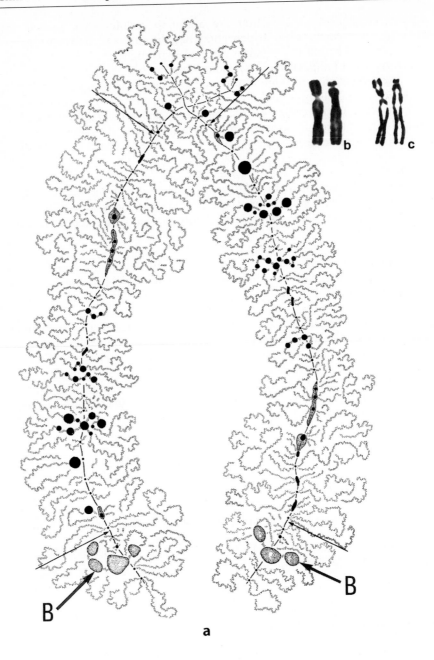

a

b

c

B

B

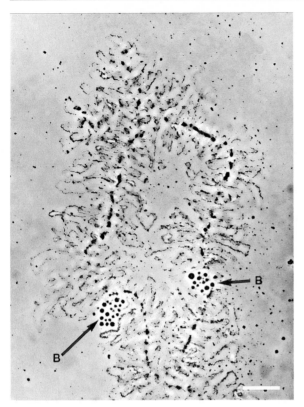

Figure V.10. Terminal region of bivalent VI carrying the pericentric inversion (micrograph J.-C. Lacroix; bar = 10 μm) (with permission of Springer-Verlag). Only the right half of the bivalent VI is represented in this micrograph, and oriented in the same way as in the schematic drawing of this inversion (Figure V.9); preparation from another oocyte of the same individual. Note the B loops localised in the lower part of the micrograph.

Figure V.9. Comparative study of a pericentric heterozygous inversion of chromosome VI, on lampbrush chromosome and on mitotic chromosome of P. waltl (ibero-lusitanian) (micrograph A. Jaylet and J.-C. Lacroix, reproduced with the permission of Springer-Verlag).

(a) Schematic drawing, on lampbrush chromosome, of the right end of the bivalent VI involved in the rearrangement. The right telomeres are at the top of the figure. The two pairs of arrows indicate the points of inversion on the chromosome on the left, and the corresponding normal segment on the chromosome on the right. The inversion points are precisely localised by the anti-parallel position of the axial chromomeres and the distribution of the loops on the two homologues. Note the B loops, characteristic of this bivalent, in the lower part of the figure.

(b) and (c) Pair of mitotic chromosomes VI, (b) without or (c) after a hypothermic treatment, to show the paracentromeric secondary constrictions. The chromosomal rearrangement observed on the mitotic chromosomes and the position of the secondary constrictions reveal that there is a pericentric inversion. Lampbrush chromosomes indicate that it is really an inversion, and permit its exact localisation. However they do not show that the inversion is pericentric, as the centromeres can not be detected on the lampbrush chromosomes of Pleurodeles.

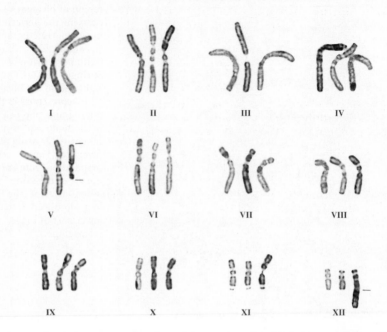

Figure V.11. Mitotic karyotype of a Pleurodeles carrier of multiple abnormalities (from Lacroix and Loones 1974, reproduced with permission of Springer-Verlag). This triploid animal belongs to a lineage which carries five reciprocal translocations detected and localised on the lampbrush chromosomes of viable adult animals. This mitotic karyotype was established from a larva of this lineage. Besides the triploidy, two translocations involving chromosomes V and XII can be observed.

V.5.1
Morphological markers

Differentiated axial structures or loop matrices correspond to different morphological types forming the families, each family having members distributed all over the karyotype. Identification of chromosomes and their mapping are based on the recognition of morphological characteristics of these families. It is therefore necessary to establish their repertory and characteristics.

V.5.1.1
Axial structures

Besides the typical chromomeres forming the axis of lampbrush chromosomes, four axial structures can be distinguished: the axial granules, dense chromomeres, *mu* structures and the structure M.

The *axial granules* can be distinguished in phase contrast by their large size. In electron microscope, they appear as double structures, formed by a chromomere and an associated protein granule; the latter does not stain with DAPI (Figure V.8a, c and d) (Pyne et al., 1989, 1995). These granules are observable in phase contrast due to their volume.

The *dense chromomeres* are rich in DNA and show a stronger DAPI fluorescence compared to other chromomeres. They are spherical and have a small size (Figure V.5b). In electron microscopy they show an ultrastructural organisation different from that of other chromomeres and as such can be easily distinguished (Figure V.8c and d). Practically all the bivalents carry these structures (Pyne et al., 1995).

The *mu structures*, located in subterminal positions on bivalents II and VIII, are excellent identification and orientation markers for these two bivalents. Even though they are of small size, they are stained with many antibodies, and thus easily identified.

The *M structure* is unique, and has been observed in only one population of Ibero-lusitanian *P. waltl*. It is localised near the sphere on the bivalent IV (Figure V.2i, j). It has been assimilated to a heterochromatic structure (Lacroix, 1968b).

V.5.1.2
Loops

The overwhelming majority of the loops do not present distinctive characteristics of size or nature of matrix, and are designated as *"normal loops"* or *N loops*. Their size varies from 20 to 30 um in length (Figures V.1a, V.5c, V.9a and V.10). Some of these loops take the form of a C loop when they are labeled with an antibody; in these loops, only one transcription unit is labeled, which extends over about a third of the loop length. The differential segment of the chromosome W carries three of this loops family. (Lacroix et al., 1990).

The other loop families are also designated by the letters A, B, C, D, P and T.

The *A loops* are giant loops having a matrix of normal appearance, which varies from 50 to 300 μm (Figure V.1b). Most of the chromosomes carry these kind of loops. Some of these loops take a different morphological aspect after labelling with an antibody. This morphological change has been observed with two polyclonal serum raised against synthetic peptides of human protein P52 (provided by Sylviane Muller, Strasbourg) (Figures V.1c, V.4, V.5a and b).

The *B loops or globular loops* carry roughly spherical or ovoid elements, each of which is formed by a transcription unit independent of its neighbours; the matrix of each unit is condensed into a compact structure upto 4 or 5 μm in diameter (Penrad-Mobayed et al., 1986) (Figure V.1a and d and Figure V.3b).

The *C loops*, or *granular loops*, are formed by granules of 0.5 to 1 mm in diameter (Figures V.1a, e and f). B and C loops make multiple associations which take median positions on the chromosomes, often near the centromeres (Figure V.1a).

The *D loops* are located near the right telomere of the bivalent II and the left telomere of the bivalent X. They appear as being formed of a small number of dense structures, or a single compact structure (Figure V.2g and Figure V.3c).

The *P loops* can only be identified in electron microscopy. These loops are formed by RNP particles having a diameter of 45 nm, whereas the other loops are formed by particles about 25 to 30 nm in diameter (Figure V.6b) (Pyne et al., 1989).

The T loops have a tubular matrix, reminding one, in phase contrast, of the structure of certain loops on the chromosomes of Drosophila spermatocytes (Hess, 1966). The loops forming this family are often observed in subtelomeric position. They have been observed in preparations from oocytes cultured for 4 to 6 days (Figure V.2h and Figure V.3a and c). Their identification in electron microscopy is also very easy because they are also formed by 45 nm particles (Figure V.6a). These loops might represent an amplification of the P loops associated with a condensation of their matrices under particular physiological conditions.

Two other types of chromosomal structures, different from loops, are the *spheres*, and the *nucleoli*.

The *spheres* are C-type snurposomes (Wu et al., 1991). Their relationship with their chromosomal organisers is to be elucidated (Figure V.1a and Figure V.2i and j).

The *nucleoli* are excellent markers of bivalent III and XI when they are present at the nucleolar organiser loci of these chromosomes, but this is very unusual in Pleurodeles.

V.5.2
Immunolabelling

Immunolabelling offers the possibility of producing a large range of new markers. Indeed, if most of the antibodies label uniformly the whole of the axis or the loops, others label only some structures. In these latter cases, the labelled structures often have an obvious relationship: they form families whose members are generally dispersed in the karyotype. Moreover, the families recognised by different antibodies can be identical, different or overlapping, on one side among themselves or, on the other side with the morphological families.

From an exploration of 350 monoclonal or polyclonal antibodies of very different origins, some forty, monoclonal or polyvalents, label positively the lampbrush chromosomes of Pleurodeles. Many of them show a comparable labelling pattern in other urodeles (Ragghianti et al., 1988). For most of these antibodies, the corresponding antigens are not known. A library of specific antibodies has not yet been established. But it is easy to form one's own library consisting of antibodies of different origins, as in most cases the labelling is perfectly reproducible. As examples, the maps presented here, show a family of loops labelled by two antibodies raised against the human protein P52, concerning the autosomes, and three families of loops differentially labeled by different antibodies, concerning the sex chromosomes (bivalent IV).

V.5.3
Mapping parameters

V.5.3.1
Chromosomes classification

The chromosomes are arranged in decreasing size, from I to XII. The lamp-brush chromosomes are however extensible structures, and can increase two-fold in length if they undergo stretching during their preparation; their relative sizes cannot therefore be used for classification of bivalents. More-over, the classification by order of relative size of lampbrush chromosomes and mitotic chromosomes do not correspond. We have therefore selected the classification of mitotic chromosomes as a reference (Lacroix, 1968a). With parallel studies of several chromosomal alterations on mitotic chromosomes and on lampbrush chromosomes, the correspondence between the two has been established. For some chromosomes, however, the orientation of the lampbrush form relative to the mitotic form is to be established.

V.5.3.2
Chromosomes orientation

The centromeres could not be localised on the lampbrush chromosomes of Pleurodeles until now. However, their position could be approximately deter-mined for most of bivalents, by comparative analysis of chromosomal rearrangements. In that case, the chromosomes are represented with the short arm on the right side.

V.5.3.3
Markers localisation

The position of chromosome markers is designated by an index corre-sponding to the distance between the left telomere and the marker relative to the total length of the chromosome. It is evaluated in hundredths of the length of each bivalent. For measuring the index, the bivalents are micro-graphed on slides in two colours, by superposing the images of the axis stained with DAPI and of the loops or markers stained with Texas Red. These slides are projected on a monitor screen of a projector. The chromosome axis and the markers are drawn on a tracing paper mounted on the screen. The measurements are taken with a map-measurer, and indexes calculated.

V.5.3.4
Correspondences between the lampbrush chromosome maps

The order and orientation of the chromosomes are the same in the three maps (Figure V.12a, b and c). Interspecific hybrids have made it possible to establish the exact correspondences between the chromosomes of the two species of Pleurodeles: *P. waltl* (Michahelles) and *P. poireti* (Gervais).

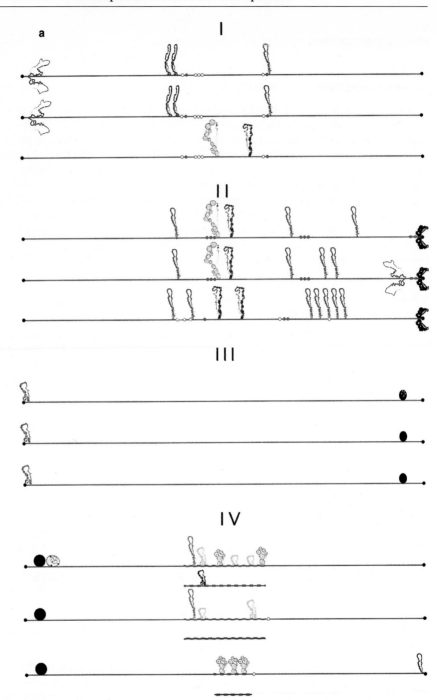

V.6
The importance of the lampbrush chromosomes

V.6.1
Detection and analysis of chromosomal rearrangements

The lampbrush chromosomes allow an accurate localisation of the chromosomal rearrangement points due to the high degree of their linear differentiation. The resolution can be as high as for the polytene chromosomes of
Diptera. Besides, the chromosomes pairing facilitates the identification of
those homologous involved in the rearrangements. Reciprocal translocations, deletions, inversions could thus be detected and analysed on these
chromosomes (Lacroix, 1968a) (Figure V.9). It is to be noted that morphological or immunological variants present in a heterozygous state on the
same bivalent represent the structural as well as functional expression of
chromosomal variations.

V.6.2
Study of populations and evolution of chromosomes

The distribution area of Pleurodeles runs from Portugal to Tunisia, across
to Spain, Morocco and Algeria. Two species are known: one, *Pleurodeles waltl*
(Michahelles) corresponds to ibero-lusitanian and Moroccan populations,
and the other, *Pleurodeles poireti* (Gervais) corresponds to the algero-
tunisian population. Individuals coming from populations situated at the
two extremities of this area are able to hybridise and give fertile interspecific offsprings (hybrids). Some markers allow the characterisation of
the two species and the observation of different geographical populations

◄————————————————————————

Figure V.12. Lampbrush chromosome maps of three populations of *Pleurodeles* (Figures
J.-C. Lacroix). Each chromosome is represented by 3 schematic drawings corresponding
respectively to the ibero-lusitanian population of *P. waltl*, Moroccan population of *P. waltl*
and algero-tunisian population of *P. poireti*.

The chromosome axis is shown in blue, the axial granules are represented by hollow
circles, the dense chromomeres by full circles. For the sex chromosomes IV, only the W
chromosome is represented in full, and for the Z chromosome only the differential
segment is represented. The differential segments are indicated by a dotted line superposed on the axis of W. The axial structures "mu" in subterminal positions on the right
of the chromosome II and on the left of the chromosome VIII are represented in red, as
they are labelled by antibodies raised against P52. They are also stained with DAPI. The
lateral structures of the maps can be identified by the schematic drawings of the Figures
V.1 and V.2. Intercalary loops A, B and C are represented by only one element of the pair.
For simplicity, the loop of the Figure V.1d has been considered as representative of B loops,
those of the Figure V.1f as representative of C loops, and those of Figure V.1c as representative of A loops, modified or not; the loops shown in colour correspond to families of
loops selectively recognised by antibodies. The nucleoli in subterminal positions on the
right of chromosome III and on the left of the chromosome XI indicate the position of
the nucleolar organisers. All the chromosomes are schematised to the same size.

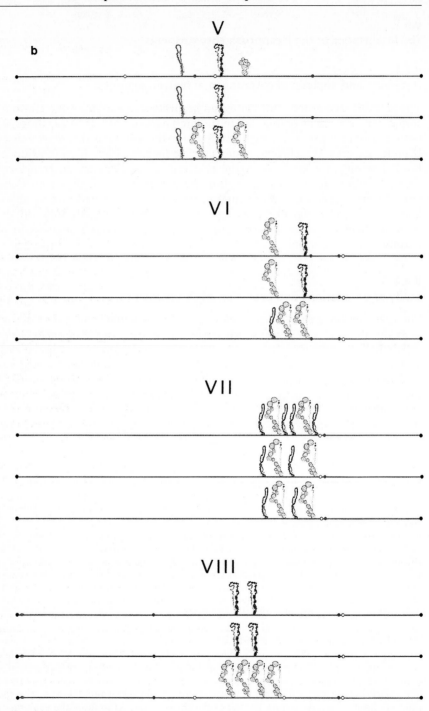

Figure V.12. *Continued.*

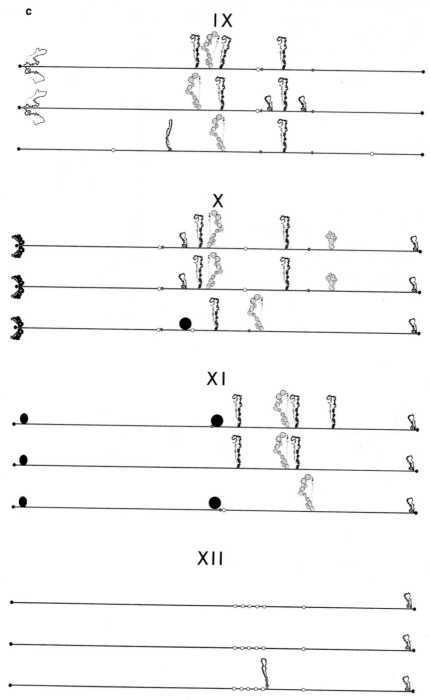

Figure V.12. *Continued.*

inside the distribution area of these species (Callan and Lloyd, 1960; Lacroix, 1968a, b).

Thus, the spheres allow us to distinguish easily the individuals of Moroccan origin (only one sphere on the bivalent IV) from those of ibero-lusitanian origin (a sphere on IV and another on XI) and of algero-tunisian origin (a sphere on each bivalent IV, X and XI).

T loops have been observed until now only in populations of *P. waltl.* One population of *P. poireti,* localised in the region of Annaba (Algeria) carries a marker on the W chromosome (Lacroix, 1970). This marker, never observed in any other population so far, was present in all the females of this population, about one hundred in all, captured in their biotope on three different stages (1966, 1967 and 1980). The M structure, absent in individuals of African origin, allowed us to establish a particular ibero-lusitanian lineage in the laboratory.

By increasing the number of chromosomal markers, it is not only possible to establish lineages identifiable by their lampbrush chromosomes but also to find individual characteristics of the animals. These observations show that the lampbrush chromosomes permit the study of the evolution of chromosomes in amphibians, at an organisational level corresponding to the DNA subdivision of the chromosome in ranges forming the loops. This organisational level is common to all the organisms, as suggested by the presence of the lampbrush chromosomes in many other groups, as well as by studies on the mitotic chromosomes and interphase chromatin.

V.6.3
Study of the transcriptional physiology *in situ*

One of the interests of the lampbrush chromosomes resides in the possibility to observe *in situ* the spectrum of particular transcription units or, on the contrary, of all the units belonging to the same family, and even the spectrum of activity of the whole karyotype. Nucleic acid and protein probes are available to permit the detection of experimentally induced cytophysiological and cytogenetic variations in the lampbrush chromosomes, with a high resolution. The giant oocytes of amphibians provides thus a particularly favourable material for genetical and physiological studies, as well as molecular and cellular.

VI
Drosophila Chromosome Study Techniques

F. Lemeunier and S. Aulard

VI.1
Mitotic chromosomes

VI.1.1
Introduction

The first description of the mitotic chromosomes of *Drosophila* was made in 1912 by Stevens. She observed four pairs of chromosomes in spermatogonia and oogonia of *Drosophila ampelophila* (synonym of *Drosophila melanogaster*). Nevertheless, the first comparative studies of *Drosophila* karyotypes were performed by Metz (1914, 1916a, b). He understood that the cytological data are of interest for evolutionary biologists. The karyotype of *Drosophila melanogaster* (2n = 8) (Figure VI.1) consists of three pairs of autosomes (pairs n°2, 3 and 4) and one pair of sex chromosomes (pair n°1). Chromosomes 2 and 3 are large metacentric, the chromosomes 4 are dot chromosomes, the X chromosome is a large acrocentric. The Y chromosome is submetacentric. The different banding techniques of mammals chromosomes may be used for the mitotic chromosomes of *Drosophila*.

The cytological analysis of the mitotic chromosomes has an important place in speciation studies, and also plays an important part for at least two other axes of specific research: a. genetic analysis of abnormal behaviour of mitotic chromosomes (more than 60 mutants were detected in *Drosophila melanogaster*) and b. cytogenetic analysis of heterochromatin. On the other hand, the cytological analysis is absolutely necessary for the identification of the rearrangements involving heterochromatin. It is difficult or impossible to detect these break points by analysis of polytene chromosomes, but they can be identified on mitotic chromosomes preparations. Nevertheless, few laboratories in the world are able to perform the techniques of *Drosophila* mitotic chromosomes analysis given their very small size.

Figure VI.1. Chromosome spread of a species belonging to the *Drosophila melanogaster* group (photographs F. Lemeunier).

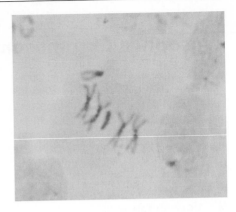

VI.1.2
Chromosome spread preparation

Principle
Tissues having a great mitotic activity are selected for the preparation of *Drosophila* mitotic chromosomes. The most convenient tissues are the neural ganglia and the imaginal discs of the third instar larvae. Before dissection the sex of the individual can be determined. The embryonic testis and ovaries appear as two white cavities located at the 1/3 back of the larva. Ovaries are smaller than testis (Figure VI.2). Mitotic preparations from the neural ganglia and polytene preparations from the salivary glands can be performed from the same larva.

There are numerous techniques of neuroblastic chromosome preparations. Some of them are described in Ashburner (1989) and Gatti et al. (1994). The technique described here was adapted from Guest and Hsu (1973).

Protocol
1 Cephalic ganglia or imaginal discs are dissected in a Ringer solution (appendix B12). The dissection can be followed by incubation in a bath of colchicine 1× (appendix V2) diluted in isotonic solution (Ringer) for 30–60 min.
2 The tissue is transferred in a hypotonic solution of 1% Na3 citrate.2H$_2$O for 10–15 min in order to swell the cell and nucleus and allow better spreading of chromosomes.
3 Fix for 3–20 min in a freshly prepared mixture of glacial acetic acid (1 vol.) and absolute methanol (3 vol.), on a depression slide covered by a cover-slip (the fixative is very volatile and can be frozen placing the slide on ice).
4 Transfer ganglia in a drop of 60% acetic acid to a separate clean slide and dilacerate using mounted needles. Do not expose material to acid any longer than necessary. Place the slide on a hot plate at 45 °C till evaporation. As an alternative, brains are dilacerated in acetic acid in a depression slide, then discarded by pipetting and transferred to a slide taking out a

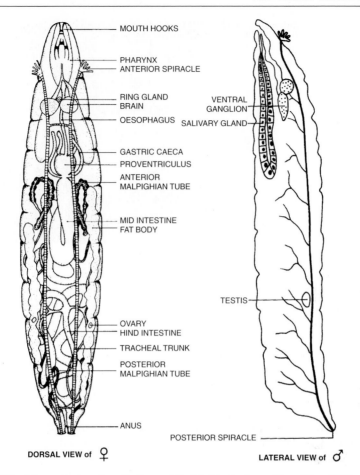

MOUTH HOOKS

PHARYNX
ANTERIOR SPIRACLE

RING GLAND
BRAIN
OESOPHAGUS

VENTRAL
GANGLION
SALIVARY GLAND

GASTRIC CAECA
PROVENTRICULUS

ANTERIOR
MALPIGHIAN TUBE

MID INTESTINE
FAT BODY

TESTIS

OVARY
HIND INTESTINE

TRACHEAL TRUNK

POSTERIOR
MALPIGHIAN TUBE

ANUS

POSTERIOR SPIRACLE

DORSAL VIEW of ♀

LATERAL VIEW of ♂

Figure VI.2. Schematic drawing of a *Drosophila* third instar larva (from King, 1965, reproduced with permission of Oxford University Press).

cold distilled water bath (the slide must be covered by a film of water) and finally air dried. Slides can be stored for some months at 4 °C for observation after a classical staining. On the contrary, fresh slides (some days only) are recommended for the chromosome banding.

VI.1.3
Staining and banding of the mitotic chromosomes: remarks

Staining and banding of the chromosomes may be performed following the techniques used for mammals chromosomes (Figure VI.3). If necessary, the slides are re-hydrated in an ethanol series (90%, 70% then 50%). The classical Giemsa staining (according to paragraph I.5.1) is routinely used to describe the general morphology of the chromosomes.

Figure VI.3. Chromosome spread of *Musca domestica*, Hoechst staining (photographs F. Lemeunier and S. Siljak-Yakovlev).

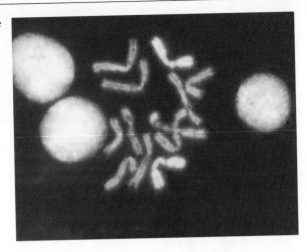

The banding techniques are: quinacrine mustard (Caspersson et al., 1968; Gatti et al., 1976; Lemeunier et al., 1978), Hoechst (Latt, 1973; Holmquist, 1975; Gatti et al., 1976), C-bands (Summer, 1972; Pimpinelli et al., 1976; Lemeunier et al., 1978), R-bands (Dutrillaux, 1975; Lemeunier et al., 1978) and N-bands (Funaki et al., 1975; Pimpinelli et al., 1976).

VI.1.4
In situ hybridisation

It is sometimes difficult to detect simultaneously the hybridisation signal and the banding pattern of the mitotic chromosomes when using radioactive or even biotinylated probes. To overcome these problems, Gatti et al. (1994) adapted the technique of fluorescent *in situ* hybridisation conjugated to a DAPI staining to produce a Hoechst banding. The fluorescence of the probe and that of DAPI are captured separately by a CCD camera, pseudo-coloured and merged on a computer. Nowadays this methodology, largely applied to human chromosomes, should be adapted to those of *Drosophila*. The first publication using these modern cytogenetic techniques belongs to Pimpinelli et al. (1995).

VI.2
Polytene chromosomes

VI.2.1
Introduction

Because of their exceptional size, lampbrush chromosomes (typical of amphibian eggs) and polytene chromosomes (e.g. salivary gland of Diptera) are also called giant chromosomes. An obvious advantage of both is their capability of more effective transcription than the normal chromosomes. Balbiani, in 1881, first discovered greatly expanded and clearly cross-striated "nuclear cords" in salivary gland cells of larvae of chironomid insects. But their significance was not realised until their rediscovery some 50 years later in Drosophilidae (Painter, 1933) and Bibionidae (Heitz and Bauer, 1933). Since then, polytene chromosomes (derived from Greek πολυ [poly, many] and τενια [tena, ribbon]) have been described in numerous insects, Collembola and Diptera. Polytene chromosomes are also found in dinoflagellates, ciliates and plants, but they do not have exactly the same kind of organisation as those of insects. Sorsa (1988a) investigates the phenomenon of polytenisation in different organisms (animals and plants) and points out the importance of the polytene chromosomes in Biology, particularly in genetic research.

In Diptera a number of larval and adult tissues, varying according to the species, possess polytene chromosomes: larval salivary glands, fat body, midgut, stomach, gastric caeca, Malpighian tubules, ring glands, nurse cells. Nevertheless, polytene chromosomes from the salivary glands offer the best cytological quality.

VI.2.2
Structure

In 1938 Geitler demonstrated that giant chromosomes are produced by endomitosis which starts at a very early stage of the ontogenesis, in cells which grow in size but do not divide. The polytenization process comprises a number of subsequent duplications in both of the homologous chromatids. The number of replication cycles, usually from nine to twelve, depends on the species and the tissue in question. The chromatids are decondensed and remain closely paired. Moreover, the somatic pairing of the paternal and maternal homologues observed in dipteran insects, allows a haploid number of chromosomes to appear in nuclei of diploid tissues. As the number of cycles increases, the chromosomes become thicker and visible in the interphase nucleus under light microscope. The higher degree of **politeny** is up to 16000 or 32000 in the salivary gland cells of certain Chironomidae (Figure IV.4). The lower one, observed in Malpighian tubules or in mid intestine of dipterous, is 32 or 64 parallel chromatids. The salivary glands chromosomes of several *Drosophila* species go through 9 rounds of replication and contain 1024 chromatids. However, the degree of polyteny is lower in the anterior part of the gland. On the contrary, when larvae were grown under optimal nutritional conditions, one more replication cycle could be observed. The length of these chromosomes is approximately 100–200 times the length of the same chromosomes in mitotic metaphase.

The polytene chromosomes present, without any preliminary treatment, an alternation of dark "bands" and light "interbands" of various width whose sequence is characteristic for a given chromosome and a given species. This natural banding corresponds, at the ultrastructural level, to the exact pairing of homologous chromomeres producing the dark bands, whereas that of the

Figure   VI.4. Polytene   chromosomes (chromosome 4) of *Chironomus thummi* (photograph F. Lemeunier).

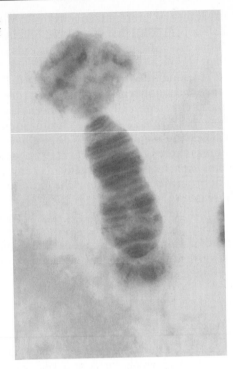

interchromomeric segments creates the light interbands. Periodically, some normally distinct bands become diffuse and swollen. These "puffs" result from the decondensation of the individual chromomeres of a band and are the sites of RNA synthesis. The result of the polytenization process is, thus, a highly reproducible pattern characteristic for a given chromosome and a certain developmental stage. The considerable interest of these chromosomes in studies of chromosomal phylogeny, chromosomal polymorphism, physical mapping, physiological genetics, comes from these characteristics.

In the polytene chromosomes of Diptera somatic pairing usually keeps the homologues tightly synapsed. Local asynapsis may appear due to technical reasons, especially too strong squashes. However, imperfect pairing may be observed in heterozygous for chromosomal rearrangements or in interspecific hybrids (Figure VI.5), because of changes in the sequence of bands.

VI.2.3
Reference maps

Genetic maps of polytene chromosomes started with salivary gland chromosomes of *Drosophila melanogaster*. The first detailed maps were drawn by C.B. Bridges (1935), then revised and completed from 1937 to 1942 by C.B. Bridges and P.N. Bridges. They were the only maps available for this species for about 40 years until King (1975) completed them with data of genes in

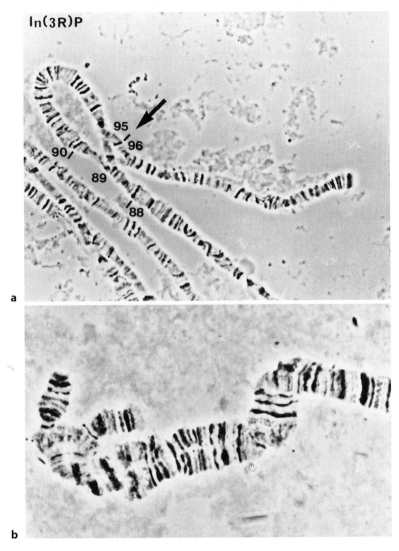

Figure VI.5. (a) Heterozygous for a chromosome inversion in *Drosophila melanogaster* (from Lemeunier and Aulard, 1992, with permission of CRC Press Inc.). (b) tip of the X chromosome in a interspecific hybrid *Drosophila melanogaster* X *Drosophila mauritiana* (photograph F. Lemeunier).

the chromosomes. Photographic maps were then published by Lefevre (1976), and Sorsa (1988b) in electron microscopy. Nowadays, polytene chromosomes of more than 100 Drosophila species are mapped. Nomenclature adopted is different from one species to another, so only that of *Drosophila melanogaster* is described.

Figure VI.6. Polytene chromosomes of *Drosophila melanogaster* (from Lemeunier and Aulard, 1992, with permission of CRC Press Inc.).

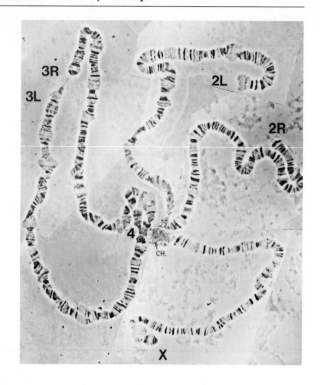

All chromosomes are joined together by their centromeric regions which undergo incomplete polytenization. There are five chromosomal arms radiating from the common chromocentre (Figure VI.6). By strong squashing, it is possible to break the centric fusion and separate individual chromosomes from the chromocentre. The complement is divided into 102 sections (1–20 for the X chromosome, 21–40 and 41–60 for the left and right arms of chromosome 2 [2L and 2R, respectively], 61–80 and 81–100 for chromosome 3 [3L et 3R, respectively], 101–102 for the dot chromosome 4). Each of the sections, starting with a prominent band, is divided into 6 subdivisions (A–F) containing a variable number of chromomeric bands (from 2 to 25) consecutively numbered from the left to the right. In this species more than 5000 chromomeric bands are observed under the light microscope. The heterochromatic Y chromosome is included into the chromocentre and, therefore, cannot be observed in the polytenic tissues.

The average DNA content per chromomere is 25 kilobase pairs. Nevertheless, because the smallest bands are composed of chromomeric units of less than 5 kb, mapping on polytene chromosomes is more precise than that on mitotic chromosomes whose banding is obtained by specific pretreatments and staining.

VI.2.4
Chromosome preparation: classical and molecular cytogenetics techniques

VI.2.4.1
Breeding conditions of Drosophila

The cytological quality of salivary glands polytene chromosomes depends not only on technical characteristics but also on the larvae themselves. In order to obtain third instar larvae of large size, they must be grown at 20 °C, on a rich medium, avoiding the excess of individuals. Thus, it is necessary to verify the quantity of eggs laid by the females. The development time of the larvae and, thus, their size and that of salivary glands, depend on the temperature. Higher temperatures, increasing the speed of the development, produce larvae, and thus, salivary glands of smaller size. At lower temperatures, larval cycle is increased and contamination of the medium may occur.

VI.2.4.2
Chromosome preparation: classical analysis

There are different techniques to prepare the chromosomes. We describe only one which is suitable for classical analysis but also for *in situ* hybridisation.

Protocol
■ **Salivary glands dissection.** If necessary, wash the larva briefly in water to remove the food medium. Larva is then transferred in a Ringer solution for insects (appendix B12). Hold the posterior end of the larva with one pair of fine forceps and with a second pair grasp the head immediately behind the black mouthparts. Pull the head gently but firmly and it should come cleanly away from the body with the paired salivary glands attached to it by the salivary ducts. The glands appear as two white elongated bags. Remove as much of the adhering fat body as possible, without damaging the glands.

■ **Chromosomes fixation.** Transfer the glands to a relative large volume of 45% acetic acid on a slide preliminary cleaned with alcohol. Leave the glands in the fixative for 1 min, but this time can vary according to the species.
  Wipe off this first fixative with a paper tissue, avoiding the glands drying and taking care not to leave paper fibres on the slide. The glands are quickly covered with a second fixative (1 vol. concentrated lactic acid – 2 vol. water – 3 vol. glacial acetic acid). Leave the glands in the fixative for 3 min. Again, this time can vary according to the species.

■ **Staining.** Remove this second fixative as the first one, with the same care, and replace with a drop of lacto-aceto orcein (according to paragraph I.5.2) and leave for 1–2 min or 20–30 min for an observation in phase contrast or light microscope, respectively. The time of this staining is empirical. It depends on the stain and the larva itself.

■ **Chromosome squash.** Blot away the excess of stain with a small quantity of liquid being left. Place a coverslip cleaned with alcohol on the glands. The slide is placed between a folded filter paper. Tap vertically over the glands with the tip of a pencil to spread the chromosomes. Then firmly press the chromosomes. Do not allow coverslip to move while squashing. This squash should be performed with the thumb or a stapler.

VI.2.4.3
Chromosome preparation: in situ hybridisation

Principle
There are different protocols for the *in situ* hybridisation, according to type of probe labelling. Preparation of the chromosomes itself depends on the authors. The protocol described here (adapted from Engels et al., 1986; Biémont and Gautier, 1988; Silber et al., 1989) is available for *in situ* hybridisation with a DNA probe (repeated or unique sequence) labelled with biotin-16-dUTP (Boehringer Mannhain) by "nick-translation" (BRL). The detection is performed using extravidin-peroxydase conjugate. The peroxydase is revealed by diaminobenzidine. The brownish reaction product, insoluble even in alcohol, can be visualised.

Protocol
■ **Dissection, fixation and squash.** The chromosomes are dissected, fixed and squashed according to paragraph VI.2.4.2. Leave the slides protected against dust for 24–48 hours at 4 °C to flatten the chromosomes. A first sorting is performed in order to store only the spreads of good quality.

■ **Coverslip removal and chromosome dehydration.** Dip the slides into liquid nitrogen for 3–4 min (or leave 20 min on dry ice) and flip off coverslip with a razor blade. Immediately place slides in 3 successive pre-cooled −20 °C ethanol baths: EtOH 50 (2 min), EtOH 75 (1 min), then EtOH 90 (1–2 min). Let them air dry at room temperature (about 2 h) protected against dust. Check slides under microscope in order to verify that the chromosomes do not become detached during the removal of the coverslip if their structure is not altered by the successive dehydration. Localisation of the squash is marked using a diamond-point pen, on the reverse side of the slide to avoid glass fragments on the chromosomes. Do not use a marker as ink is diluted with alcohol. Slides can be stored at 4 °C for one month, maximum.

■ **Pre-treatment for in situ hybridisation.** One tank of 2× SSC (appendix B4), two tanks of 70% EtOH, and one tank of 90% EtOH are placed in 65 °C water bath. Heat slides in 65 °C 2× SSC for 30 min. At the end of 30 min, take all the tanks out of the water bath and rinse the slides in 70% EtOH tanks, each for 5 min, then 5 min in 90% EtOH. In this way, the slides return gradually to room temperature. Let them air dry in dust-free atmosphere at room temperature. At this stage, another slides selection is performed. They can be stored at 4 °C for months.

VI.2.5
In situ hybridisation

Protocol
■ **Chromosome denaturation.** Slides are immersed for 1.5 min in 0.07N NaOH (dilution freshly prepared from a stock solution of 7N NaOH stored at 4 °C). Rinse three times for 5 min each in 2× SSC (appendix B4). Dehydrate in 70% EtOH, two times for 5 min each, then in 90% EtOH, one time for 5 min. Let slides air dry at room temperature.

■ **Hybridisation.** Heat the hybridisation solution containing 1 µg of labelled DNA probe for 5–10 min at 100 °C and place immediately in ice. Put 10 µl of the probe solution on the squash, cover with a clean coverslip and seal with rubber cement. Put the slides horizontally in a moist chamber (hermetic plastic box with a layer of filter paper soaked with 2× SSC (appendix T4) on the bottom and incubate overnight in water bath at 37 °C.

■ **Posthybridisation and detection (biotinylated probe).** Remove the rubber cement with forceps. Float off coverslip in pre-warmed 2× SSC if it remains on the slide. Wash twice for 10 min each in 2× SSC pre-warmed to the hybridisation temperature, then one time for 2 min in a PBS solution (appendix B1) containing 0.1% Triton ×100 pre-warmed at 30 °C and, at last, two times for 10 min each in 1× PBS at room temperature. Slides are drained just for the time needed to add extravidin to the solution of BSA (appendix H8). Do not let them dry. Put 50 µl of solution on the squash and cover with a large coverslip. The slides are incubated horizontally in a moist chamber for 30 min at 37 °C. Rinse the slides in a PBS solution containing 0.1% Triton ×100 for maximum 1 min, at room temperature, then 2 times for 5 min each in 1× PBS. Slides are drained for just the time needed to add the H_2O_2 to the solution of DAB (appendix H9). This stain must be freshly prepared. Cover the squash with the DAB solution. Leave the slides horizontally (with or without coverslip) protected against light at room temperature for 5–10 min. (**Attention:** taking into consideration the toxicity of DAB, it is necessary to handle DAB solution with gloves and treat excess solutions and contaminated material with household bleach during cleanup).

■ **Chromosomes staining.** Slides are drained over a container of household bleach and rinsed quickly in a first bath of 1× PBS (appendix B1) then 10 min minimum in a second bath of 1× PBS. Slides can be kept in PBS for several hours until ready for staining. Stain chromosomes in Giemsa (according to

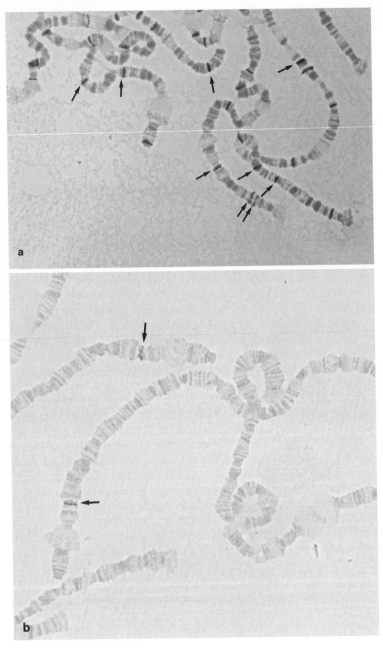

Figure VI.7. *In situ* hybridisation of a biotinylated probe on the polytene chromosomes of *Drosophila melanogaster*. (a) the 412 transposable element. (b) *white* gene (photographs F. Lemeunier and S. Aulard).

paragraph I.5.1) for 3 min, rinse quickly in tap water and air dry. Mount in Entellan.

VI.2.6
Chromosome observation

Chromosomes are observed under a photomicroscope in phase contrast with oil lens. Photographs allow the identification of chromosomal rearrangements (Figure VI.5) by comparing with the reference maps, or the localisation of the hybridisation signal (Figure VI.7).

VII
Techniques of Interphasic Nucleus Study

VII.1
Sex chromatin examination

P. POPESCU

Principle
In most Mammals species having a XX/XY sex determination one of the X chromosomes is inactivated in the female cells early in the blastocite stage. According to Lyon theory, the inactivation of one X chromosome in female cells is a mechanism of dosage compensation for the X linked genes having a different dosage in XX females and XY males. The inactivated X chromosome remains in a condensed state through the interphase and is easily visible in a variable cells proportion after staining. It is called "Barr body" or sex chromatin body. For routine clinical investigations, the buccal smear is the most easily obtained and common source of cells. There are different techniques of sex chromatin analysis. The technique described herein is based upon the Moore and Barr method (1955).

Protocol
Cells of the buccal mucosa are scraped with a metallic spatula or a glass slide. The material is spread over a clean slide and fixed immediately placing the slide in fixative solution (3 vol. ethanol – 1 vol. acetic acid), for 30 minutes. The preparation is hydrolysed in a HCl 5 N solution at room temperature for 20 minutes, then rinsed in tap water for 3 minutes. The slide is stained with an aqueous solution of 0.5% (weight/vol.) creysle violet or 0.1 (weight/vol.) toluidine blue, for 5 minutes.

Sex chromatin is seen as a dark blue body at the periphery of the nuclear membrane. This body is seen in a fraction of the total observed cells: 5 to 30% of the epithelial cells of buccal smear and 70% to 90% of fibroblast cell cultures.

Acridine orange staining (according to paragraph I.5.3) gives a better differentiation between the sex chromatin body and the rest of the chromatin (Figure VII.1).

Figure VII.1. Sex chromatin body (indicated by an arrow) in women interphase nuclei, stained with acridin orange (photographs B. Dutrillaux).

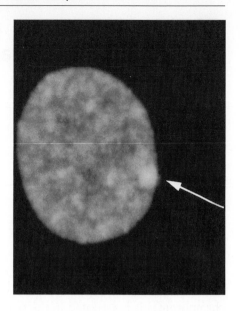

Remarks

The number of Barr bodies observed in a cell is equal to the number of its X chromosomes less the active X chromosome. It is an easy and rapid technique but at least 200 to 300 cells should be examined for a correct analysis.

VII.2
Released chromatin (Chromatin halo)

P. Popescu

In situ hybridisation on released chromatin (chromatin halo) is an important tool for the gene mapping as it makes it possible to observe the position of DNA sequences very closely on DNA fibre.

The preparation of released chromatin or DNA fibres consists in the elimination of the nuclear proteins in the interphasic nuclei using a saline treatment and/or a detergent and the suspension spreading on slides. The decondensed chromatin spreads as DNA loops, from the chromosome skeleton and appears like a halo. There are different techniques described in literature (Weier et al., 1995), to improve the de-condensation homogeneity and the linearity of the DNA fibres.

Protocol

The protocol is described by Parra and Windle (1993).

A fibroblasts or lymphocytes suspension obtained by classical techniques is centrifuged at 150 g for 5 minutes, to concentrate the cells. After the counting, the cells are re suspended in a PBS⁻ solution (appendix B1) at a final concentration of 50 to 2500 cells/µl. A drop of cell suspension diluted in this manner is placed on the end of a clean slide and allowed to dry in the open air at room temperature. The cells are then immediately lysed: a drop (5 to 6 µl) of a buffer solution Tris pH 7.4 at 200 mM (dilute stock solution at 1M-appendix S14), containing 0.5% SDS (weight/vol.) and 50 mM EDTA (appendix S4). After 7 to 10 minutes of incubation in a humid chamber at room temperature, the slide is placed on a support in vertical position as the drop of DNA in suspension flows out along the slide. The slide is allowed to dry in the open air, then immersed in the fixative (3 vol. methanol-1 vol. acetic acid) for 5 minutes at 4 °C. The DNA beam is observable under the microscope and must be without nuclei.

In situ hybridisation is performed on these slides according to usual techniques.

VII.3
In situ hybridisation of interphasic nuclei

C. BURGEOIS

Principle
In situ hybridisation of DNA probes of interphasic nuclei and their revelation by immunofluorescence (Cremer et al., 1986), and of metaphase chromosome spreads are based upon the same principles.

The difficulties concerning in situ hybridisation of interphasic nuclei consist in:

- probe penetration to hybridisation site,
- DNA denaturation of interphasic nuclei.
- and particularly reading of results. The hybridisation signals are placed in a three-dimensional space as the nuclei represents a volume.

Protocol
The different points described above are detailed, taking into consideration the type of preparation used for the nuclei revelation. There are several methods: paraffin sections, frozen sections, whole cells spreads or isolated nuclei. If the treatments are aggressive, material fixed on "super-frost" or special treated slides are recommended to avoid the material loosening during the in situ hybridisation treatments.

■ **Techniques increasing the probe penetration.** There are different protocols: sections in paraffin, nuclei and cells prepared by classic spreads or slide with cell layers.

Protocol for paraffin sections
A pre treatment using paraffin sections could be realised in two steps (deparaffin the section and facilitate the probe penetration). The sequences of

Figure VII.2. In situ hybridisation to a human interphasic nuclei of a specific probe of the Y human chromosome, biotinylated (Boehringer Mannheim) and visualised by streptavidine Texas red-conjugated. Counterstaining by DAPI (photographs D. Celeda).

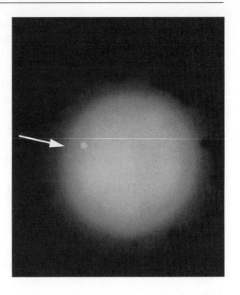

the steps should be adapted according to the examined material, in this case a 5 μm thick-thyroid section.

1 The section is deparaffined as follows: the slide is incubated in three successive xylene baths for 5 minutes each, followed by two 100% ethanol baths for 4 minutes each and in a bath of PBS⁻ buffer solution (appendix B1) for 10 minutes.

2 To make the probe penetration easier, the sections are incubated in a bath of SSC2× pH7 buffer solution (appendix B4) at 70 °C for 20 minutes. The sections are then incubated in a bath of pepsin solution of 4 μg/ml HCl_2N at 37 °C for 8 to 12 minutes. The pepsin action is stopped by two successive washes for 5 minutes in bi-distilled water and two baths of buffer solution SSC2× pH 7 at room temperature.

These two steps are followed by a post-fixation for 10 minutes in a buffer solution PBS⁻ containing 50 mM $MgCl_2$ (dilute stock solution at 1M-appendix SS9) and 1% (vol./vol.) formaldehyde. The classical treatments of *in situ* hybridisation of chromosomes are then performed.

Protocol for the nuclei and spread cells or slides with cell layer

A proteolytic treatment using pepsin or proteinase K is necessary. The slide is re-hydrated in a phosphate buffer solution (appendix B10) for 5 minutes, then it is incubated in a bath containing 50 μg/ml HCl 0.01 N pepsin solution for 10 minutes at 37 °C or 0.5 μg/ml protease K solution in PBS⁻ (appendix B1) for 10 minutes at 37 °C. This treatment is performed either before or after the digestion by ribonuclease A of the *in situ* hybridisation of chromosomes (paragraph III.2.3).

■ **Denaturation of interphasic DNA.** The denaturation time depends on the examined material. Generally, it is about 3 minutes (2 min. minimum and 5 min. maximum) and it is performed in a SSC2× pH 7 solution (appendix B4) containing 70% de-ionised formamid (appendix M7) at 70 °C. The minimum time of denaturation for the tissues sections is 3 min.

■ **Hybridisation and probe revelation.** The usual protocols of in situ hybridisation as described in Chapter III are followed by other steps.

■ **Observation and registration of the results.** The observation and registration of hybridisation signals depend very much on the observation of the depth of focus. If it passes beyond 3–4 μm, the best solution is to use a fluorochrome where the excitation wavelength is not lying upon that of the fluorochrome used for the probe detection. e.g., if the probes are detected by the rhodalgreen or fluorescein rhodamin or texas-green conjugated, the DAPI should be used for the counter-staining of the DNA nuclei (Figure VII.2).

Remarks
The results obtained in the laboratory show that the fixative (acetic Carnoy, paraformaldehyde, ethanol, acetone or another fixative) is not a very important factor for the quality of the results if the pre-treatments are well adapted.

VIII
Application of Flow Cytometry and Slit-Scan Fluorometry in Mammalian Chromosomes Analysis and Sorting

M. Hausmann and C. Cremer

VIII.1
Introduction

Fluorescence activated analysis and sorting ("flow cytometry") offers a successful method to characterise particles in a fast, statistically quantitative way (Melamed et al., 1990). Originally developed to analyse and sort cells (Bonner et al., 1972) it is nowadays a highly versatile tool for biomedical research and routine clinical diagnostics (Darzynkiewicz et Crissman, 1990). Flow cytometry was introduced into cytogenetics about 20 years ago (Gray et al., 1975; Stubblefield et al., 1975). "Flow cytogenetics" (Gray and Langlois, 1986; Cremer et al., 1989a) contributed significantly to medicine and molecular biology.

Consequently, flow cytometric chromosome classification ("flow karyotyping") has proved to be useful for the chromosomes analysis of several species, e.g. Chinese hamster (Bartholdi et al., 1984; Barths, 1987; Otto and Tsou, 1985), muntjac (Carrano et al., 1976; Lee et al., 1994), rat kangaroo (Stöhr et al., 1980), chicken (Stubbefeld et Oro, 1982), mouse (Dietesche et al., 1981; Dixon et al., 1992), wood lemming (Hausmann, 1984), swine (Dixon et al., 1992; Grünwald et al., 1986; Blaise et al., 1990) bovine (Dixon et al., 1992; Schmitz et al., 1995), and mammalian hybrid cells (Lee et al., 1994; Dixon et al., 1992; Cremer et al., 1982; van Dilla and Deaven, 1990). Flow karyotyping was extensively applied to normal and aberrant human cell lines (Carrano et al., 1979; Langlois et al., 1982; Harris et al., 1987; Trask et al., 1989; Boschmann et al., 1991). High purity sorting of large amounts of chromosomes (Cremer et al., 1984) allowed the construction of DNA libraries for all types of human chromosome (van Dilla et Deaven, 1990).

Conventional flow karyotyping only registers integrated values (1–2 fluorescence parameter, scatter light). This technique allows the analysis of the chromosomes according to their DNA content and/or DNA base composition. More detailed morphological information can be obtained by slit-scan flow fluorometry (cytometry) (Wheeless, 1990) which was derived from automated cytopathology screening in microscopy (Wheeless and Patten, 1973). Current slit-scan systems (e.g. Hausmann et al., 1992) provided chromosome features such as the number (Lucas et al., 1991; Rens et al., 1993) and relative positions of centromeres (Hausmann et al., 1992; Lucas et al., 1993; Gray et al., 1979; Lucas et Gray, 1987; Boschmann et al., 1990; Rens et al., 1994; Hausmann et al., 1993) of banding pattern (Bartholdi et al., 1989). Two parameter slit-scan flow fluorometry was performed to detect specifically labelled human chromosomes and interspecies translocations in Chinese hamster/ human hybrid cells (Hausmann et al., 1991) after fluorescence in situ hybridisation (Cremer and Cremer, 1992) in suspension (Dudin et al., 1987), human chromosomes and chromosome translocations inter-species present in human × hamster hybrid cells (Hausmann et al., 1991).

VIII.2
Principle of the flow cytometry

The following text will concentrate on three axial, laser based systems. Besides those the application of microscope based, epi-illuminated flow cytometers (coaxial flow cytometers) was published (Dittrich et Göhde, 1969; Otto et Tsou, 1985).

VIII.2.1
Standard flow cytometry

In its least complex form flow karyotyping and sorting is performed with a one laser device (Figure VIII.1) using one or several detection parameters (corresponding to one or several colours of chromosome labelling).

Isolated, fluorescence stained chromosomes are injected coaxially into a laminar streaming sheath buffer- in a flow cell. By means of hydrodynamic focusing through a jewel orifice of typically 50–80 μm diameter at the exit of the flow cell, the chromosomes are confined and aligned within the core of this stream ("jet-in-air" device). They pass one by one the laser beam that excites the stained chromosomes to fluoresce. For a flow velocity of about 10 m/s, typical particle rates are in the order of several hundreds up to a thousand per second. In more complex systems two or three lasers (mostly argon-, krypton-ion, or dye-lasers) exciting the chromosome fluorescence sequentially are used (e.g. Lebo et al., 1987).

The fluorescence emission of each chromosome is detected by a photo-multiplier (PMT) during the few microseconds the particle takes to traverse the focal volume. The axes of the excitation optics, the detection optics and the fluid jet are perpendicular to each other (three axial flow cytometers). In case of chromosome labelling by more than one dye (e.g. Hoechst 33258 and Chromomicyn), A3 representing the ratio of AT to GC base pairs, the emission of each dye is detected by a separate photomultiplier (PMT). Cross talk between the detection channels can be overcome by appropriate light filters. The PMT signal of each particle is usually either linearly or logarithmically integrated, digitised, and registered in a multi-channel analyser (typically 256 or 512 channels). This results in a histogram (particle number vs. intensity), in which, ideally, each chromosome type is represented by an isolated peak.

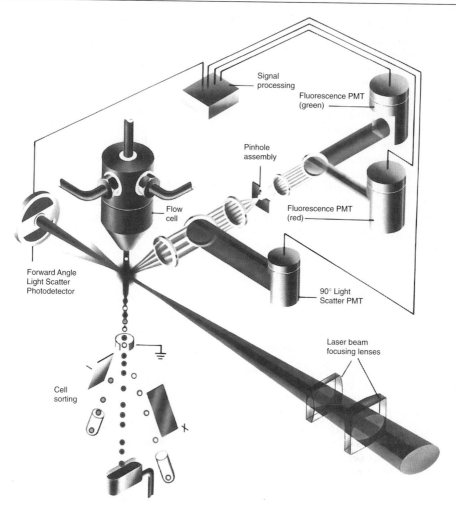

Figure VIII.1. Schematic representation of a three-axial flow cytometer (from Cremer et al., 1989b, reproduced with the permission of Berichte der Bunsen-Gesellschaft für Physikalische Chemie).

This peak pattern is typical for a given cell line, such histograms are also called "flow karyotype".

VIII.2.2
Slit-scan flow fluorometry

In cases where the integrated fluorescence of a chromosome does not give enough information for the discrimination, a further parameter may be an assessment of the fluorochrome distribution along the chromosome (i.e. information about the morphology). Due to the chromosome alignment parallel to the fluid jet and thus perpendicular to the laser beam, a spatially resolved fluorescence excitation can be obtained by highly focusing the laser beam to an ideal ribbon shape (Figure VIII.2). Although the complete optics for slit-scanning is not commercially available from the manufacturers of flow cytometers, it can be designed by on the shelf components.

Such slit-scan systems have a practical resolution of about 1–2 μm (Rens et al., 1994; Hausmann et al., 1996b) which appears to be sufficient for chromosome analysis using preparation techniques that increase the chromosome length. Thus, with an appropriate fast electronics the registered intensity distribution of each particle (and each fluorochrome) can be shown by a profile (relative intensity, – vs. time of flight of the particle). For DNA specific dyes such a profile shows a considerable intensity reduction in the centromere region so that the centromeric index can be determined (Hausmann et al., 1993).

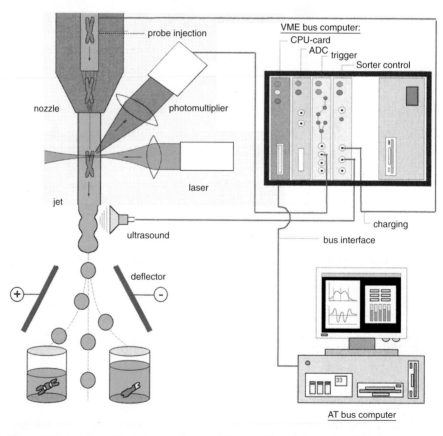

Figure VIII.2. Schematic representation of a "slit-scan" flow fluorometer (from Hausmann et al., 1996a, reproduced with the permission of Zeitschrift für Medizinische Physik).

VIII.2.3
Flow sorting

The advantages of flow sorting are that the analysis is performed for each individual particle and this analysis is independent from sorting. This means that the sorting parameters can be freely chosen. For sorting, the fluid jet of a three axial system is modulated in diameter by a piezo crystal driven vibration of the nozzle. This results in droplet generation at a defined position (typically $500\,\mu s$ – $1\,ms$ after particle detection). These droplets contain the chromosomes. If a chromosome fulfils the sorting criteria, the respective droplet is electrically charged via the flow jet before it is fully separated. By means of a high voltage electric field, which all droplets pass, negatively and positively charged droplets are deviated from non-charged ones and collected in different sorting vessels.

VIII.2.4
Computing

Recent development on the computer market has offered the possibility to fulfil the requirements of flow cytometry control and data analysis by appropriate configurations of personal computers (PC). Most manufacturers offer their flow cytometers with computing systems on PC basis including complete software for measurement control, data acquisition and evaluation of histograms. Unfortunately, the commercially available algorithms for histogram analysis are usually optimised for the analysis of immuno-labelled cells or cell cycle analysis by DNA content. Thus, for the quantitative analysis of flow, karyotypes home-made programs might be helpful (Dölle et al., 1991; van den Engh et al., 1990).

Slit-scanning, especially slit-scan sorting requires real time computing. The acquired staining pattern (profile) of each particle has to be evaluated for sorting decision in less than $1\,ms$ (time between analysis and droplet break off). This can be achieved by analogue systems (van den Oven et Aten, 1990) but with the disadvantage of being limited in the choice of sorting parameters.

Figure VIII.3 shows the computer system of the Heidelberg slit-scan fluorometer (Hausmann et al., 1996a). It consists of a VMEbus system for real time computing and a PC/AT based system on which the operations are performed which are not critical in run time. The system evaluates slit-scan profiles in less than $600\,\mu s$ according to centromere number and position.

VIII.3
Methods of chromosome preparation and staining

There are as many or even more protocols or protocol modifications published than species being analysed by flow cytometry. The following examples can, therefore, neither be understood to be complete nor to be the best

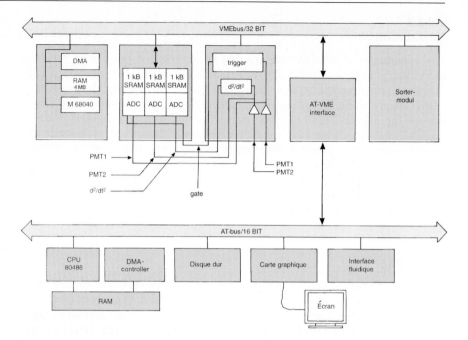

Figure VIII.3. Configuration of the computer system of the Heidelberg slit-scan fluorometer (from Hausmann et al., reproduced with the permission of SPIE – Optical Engineering; figure J. Dölle, 1994).

for all kinds of species. The selection depends more or less on personal experience of the authors for individual cases of cell lines from different species (human, Chinese hamster, pig, wood lemming, Chinese hamster × human hybrid cell lines).

VIII.3.1
Modified Hexanediole method

References: Barths (1987); Stöhr et al. (1980); Dudin et al. (1987).

Protocol
The cells growth on Ham's F10-FCS 10% (appendix CM1) containing 1% hypoxanthine (final concentration 100 μM) and thymidine (final concentration: 16 μM). Accumulation of mitosis cells is produced adding 8 h colcemid at the final concentration of 0.25 μg/ml, 8 hours before harvest. The mitotic cells are harvested by shake off, transferred into a tube and centrifuged at 350 g for 15 min. For the hypotonic treatment, the pellet is successively incubated for 3 min at –20 °C, 10 min at +4 °C, 1 ml pre-warmed hypotonic solution Tris-NaCI-MgCl$_2$ (appendix F1) is added, and incubated for 15–25 min at room temperature. The cells are again centrifuged, at 350 g, for 10 min., re-

suspended in 1 ml hexanediol solution (appendix F2) and incubated for 10 minutes at room temperature. For chromosome isolation, the cells are "blown out" either by sonication or syringing (22 G needle). The chromosomes are stained either by DAPI staining (final concentration: 1.75 µg/ml, appendix S2) for 15 minutes, or by ethidium bromide staining (final concentration: 25 µg/ml, appendix SS3) or by dual staining with Hoechst 33258 (final concentration: 2 µg/ml, appendix S6) and chromomycin A3 (final concentration: 40 µg/ml, appendix S16) for about 2 hours.

Remarks
Chromosomes can be stored at 4 °C for a long time (up to 4 years) in hexanediol solution. The technique is well suited for FISH in suspension.

VIII.3.2
TAcCaM – method

References: Hausmann et al. (1993).

Protocol
Cells are cultured, according to the paragraph VIII.3.1, in a Ham F10-FCS 10% medium (appendix CM1) containing hypoxanthine (final concentration 100 µM) and thymidin (final concentration 16 µM). Accumulation of the mitotic cells is produced by colcemid block at a final concentration of 10 µg/ml 5 hours before harvesting. The cells are harvested by shaking off, transferred to a tube and centrifuged for 10 min. at 350 g. For the hypotonic treatment the pellet is incubated in 1 ml of 7.5 mM KCL solution (appendix M1) for 5 minutes at room temperature. The cells are centrifuged for 10 min at 350 g, then resuspended in 1 ml of buffer TAcCaM (appendix F3), incubated for 10 min at room temperature and sonicated or syringed (22 G needle), to isolate the chromosomes. The isolated chromosomes are stained: a) with DAPI (concentration: 1.75 µg/ml, appendix S2) for 15 minutes, or b) with propidium iodide (final concentration 1 µg/ml, appendix S12) for 15 min or c) with ethidium bromide (final concentration 25 µM, appendix S3).

Remarks
The chromosome suspension can be stored for several months at 4 °C. The preparation method is well suited for slit-scan flow fluorometry of pig chromosomes. TAcCaM buffer solution preserves the chromosome morphology and allows the GTG banding (according to paragraph VIII.4.5) on sorted chromosomes (Hausmann et al., 1993).

VIII.3.3
Methanol – Acetic Acid method

References: Celeda (1993).

Protocol
Cells are cultured and harvested according to the paragraph VIII.3.1, but the centrifugation is performed at 150 g instead of 350 g. For the hypotonic treatment: the pellet is incubated for 5 min at −20 °C, 10 minutes at 4 °C, incubated again in 1 ml of KCL 75 mM (appendix M1) for 20 minutes at room temperature.

Cells are centrifuged 10 min at 350 g, resuspended in 1 ml of a mixture 3 vol. methanol-1 vol. acetic acid at 4 °C and sonicated or syringed (22 G needle). The isolated chromosomes are DAPI stained (final concentration: 1.75 μg/ml, appendix S2), for 15 min.

Remarks
This method has been applied for FISH in suspension of human lymphocytes. Chromosomes of Chinese hamster appeared highly condensed but could be stored for several months at 4 °C.

VIII.3.4
Tris/MgCl$_2$,/Triton X-100 method

References: Otta et al. (1980).

Protocol
The cell are cultured according to the paragraph VIII.3.1. For the hypotonic treatment the cells are incubated for 45 min on ice and centrifuged for 10 min at 150 g.

The pellet is resuspended in 1 ml hypotonic solution Tris-MgCl$_2$ BET-MA-RNase (appendix F4) and incubated for 10 min at room temperature. 50 μl of a 10% solution Triton X-100 is added, the cells are again incubated for 10 min at room temperature and syringed with 22 G needle. The isolated chromosomes are stained with ethidium bromide adding 1 ml staining solution Tris-MgCl$_2$-NACI-BET-MA (appendix F5).

VIII.3.5
Polyamine method

References: Blumenthal et al. (1979); Lalande et al. (1984).

Protocol
The cells are cultured and harvested according to paragraph VIII.3.1, except that centrifugation is at 150 g instead of 350 g. For the hypotonic treatment cells are washed once in a buffer BSS$^+$ without NaHCO$_3$ (appendix B2) con-

taining HEPES at 25 mM final concentration (dilution of stock solution at 1 M appendix SS2) centrifuged for 10 min at 150 g, resuspended in 1 ml solution KCI 75 mM (appendix M1) and incubated for 25 min at room temperature. The pellet is centrifuged for 8 min at 150 g, washed in 1 ml P1 (appendix F6), centrifuged and resuspended in 1 ml chromosome isolation buffer P2 (appendix F7). The cells are sonicated for 1–2 min. The chromosomes are DAPI stained (final concentration: 1.75 µg/ml, appendix S2), for 15 min.

VIII.3.6
Modified Polyamine method

References: Sillar et Young (1981); Boschman et al. (1991).

Protocol
The cell are cultured in MEM-FCS 10% (appendix CM1) with non essential amino acids 1% (final concentration). The mitotic cells are blocked with vindesin sulphate block (0.35 µg/100 ml) for 3 hours before harvest. The mitotic cells are harvested by shaking off, transferred to a tube and centrifuged for 5 min at 200 g. For the hypotonic treatment, the pellet is resuspended in 1 ml KCI at 37.5 mM (appendix M1) and incubated for 15 min at 37 °C. The cells are centrifuged for 5 min. at 200 g, washed in 9 ml PM1 (appendix F8), centrifuged for 5 min at 200 g, then resuspended in 1 ml buffer PM2 (appendix F9) and incubated for 15 min at 37 °C. The cells are syringed with a 22 G needle. For the staining, 20 µl of $MgSO_4$ at 0.5 M (dilution of stock solution at 1 M, appendix SS10), then Hoechst 33258 (final concentration: 1 µg/ml, appendix S6) and chromomycin A3 (final concentration: 50 µg/ml, appendix S16) are added to chromosome suspension which is incubated for 90 min at room temperature. Add Na_3 citrate $2H_2O$ (dilution of stock solution at 0.1 M, appendix SS15) at a final concentration of 10 mM, the suspension is incubated again for 30 min at room temperature.

VIII.3.7
Hepes/MgSO$_4$ method

References: van der Engh et al. (1984).

Protocol
The cells are cultured and harvested according to paragraph VIII.3.1. For the hypotonic treatment, the cells are incubated for 10 min at −20 °C, centrifuged for 8 min at 150 g and resuspended in 1 ml hypotonic solution HEPES-$MgSO_4$-KCL-DTT-RNase-PI (appendix F10). The cells are incubated for 10 min at 37 °C, a solution of 100 µl Triton ×100 is added and the cells are incubated again for 10 min at room temperature.

The cells are syringed (22 G needle) or sonicated and incubated again for 30 min at 37 °C, then stained with DAPI (final concentration: 1.75 µg/ml, appendix S2).

VIII.3.8
Modified Hepes method

References: B. Hagman, M. Hausmann, C. Cremer (unpublished).

Protocol
The cells are cultured according to paragraph VIII.3.1. The mitotic cells are blocked adding 5 µg/100 ml colcemid (final concentration) and incubated for 4 hours before harvest. The mitotic cells are harvested by mechanic shock and centrifuged for 10 min at 150 g. For the hypotonic treatment, the pellet is resuspended in 400 µl hypotonic solution KCl-HEPES (appendix F11) and incubated for 10 min at room temperature. Then, 200 µl solution KCl-HEPES-TX100-P1 (appendix F12) is carefully added and the cells are incubated for 10 min at room temperature. The cells are syringed with 22 G needle. For staining with propidium iodide the chromosome suspension is diluted 1/10 with a solution KCl-HEPES-PI (appendix F13). DAPI staining (according to paragraph VIII.3.7) is also possible.

Remarks
The unstained chromosome suspension can be stored at 4 °C for several days. The method is well suited for flow karyotyping and slit-scan flow fluorometry of chromosomes of Chinese hamster cell lines.

VIII.3.9
FITC labelling by in situ hybridisation suspension

References: Dudin et al. (1987).

Protocol
The chromosome suspension is prepared according to VIII.3.1. They are centrifuged for 15 min at 350 g, washed in 1 ml 2× SSC buffer pH 7 (appendix B4), centrifuged for 15 minutes at 350 g, and resuspended in 0.5 ml of hybridisation mixture for chromosome in suspension (appendix H6) containing 1 µg of the biotinylated DNA probe. The chromosome DNA is denatured at 73 °C for 6 minutes and renatured (hybridisation) at 38 °C–42 °C for one night. The chromosome suspension is centrifuged for 10 min at 350 g, washed in 1 ml of 2× SSC buffer pH 7 (appendix B4), prewarmed at 38 °C, centrifuged, washed in 1 ml 0.1× SSC buffer pH 7, centrifuged, washed in 1 ml of 2× SSC buffer pH 7, centrifuged and finally resuspended in 1 ml of KCL-HEPES-MgSO$_4$ (appendix F14).

The hybridisation signals are detected by FITC double antibody labelling according to manufacturer's protocol (paragraph III.2.8). The chromosome suspension is stored for 30 hours at 4 °C. Then it is counterstained with DAPI (final concentration: 1.75 µg/ml, appendix S2) for 15 min or propidium iodide (final concentration: 1 µg/ml, appendix S12) for 15 min and antifading PPD8 (appendix S11) is added.

Table VIII.1. List of staining solutions regularly used for flow cytometry and their characteristics and corresponding lasers

Staining solution	Excitation (nm)	Emission (nm)	Specificity	Laser
DAPI	310–380	400–530	base pair AT	Ar, Kr (U.V.)
DIPI	300–400	400–530	base pair AT	Ar, Kr (U.V.)
Hoechst 33258	310–380	400–530	base pair AT	Ar, Kr (U.V.)
chromomycine A3	350–460	>460	base pair GC	Ar, Kr (visible)
mithramycine A	350–460	>460	base pair GC	Ar, Kr (visible)
ethidium bromide	300–400 440–570	>560	no specific	Ar (visible)
propidium iodide	300–400 440–570	>560	no specific	Ar (visible)
FITC	440–520	480–560	(antibody)	Ar (visible)

Remarks

For slit-scanning (Hausmann et al., 1991) KCl-HEPES-MgSO$_4$ (appendix F14) is added to obtain an appropriate chromosome concentration. For stabilisation of the chromosomal morphology, absolute ethanol is added (30% of the final suspension volume is added drop-wise under continuous pipetting)

Dyes and lasers

In Table VIII.1, the dyes usually applied in flow cytometry and the appropriate lasers are summarised.

VIII.4
Measurements and evaluation in flow cytometry

VIII.4.1
Flow karyotypes

The preparation techniques described above cannot be used arbitrarily for any given cell type. Moreover the quality of a flow karyotype (i.e. the resolution in significant peaks) is significantly influenced by the combination of preparation and staining technique which should be optimised for the cell type to be measured and its cell culture conditions. A metaphase spreads looking "nicely" under the microscope observation cannot be concluded to obtain an optimal flow karyotype.

Figure VIII.4 shows examples of flow karyotypes from the Chinese hamster cell line CHL prepared according to four different protocols. The microscopically visible morphology of the chromosomes was quite reasonable in all the cases. The flow karyotypes, however, differ significantly. These flow karyotypes may be an extreme example and should not be generalised for all cell types.

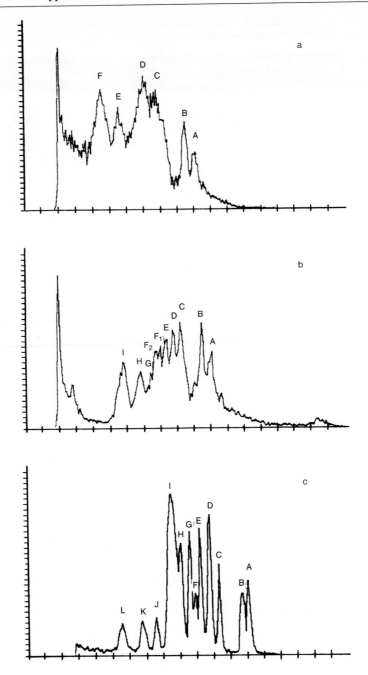

To obtain optimally resolved flow karyotypes they can be registered on linear or logarithm scale. This means that the input values (PMT signal) are either linearly or logarithmically distributed on the histogram channels (typically 256). Figure VIII.5 shows this distribution for an EPICS V flow cytometer and two corresponding flow karyotypes of the same chromosome preparation.

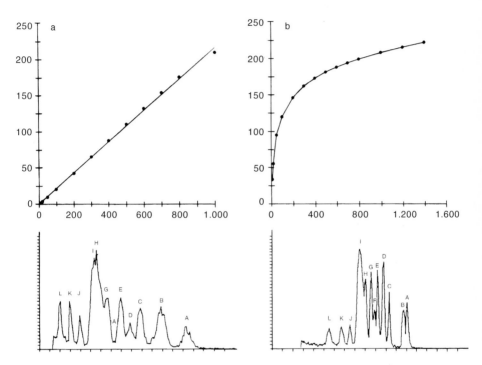

Figure VIII.5. Histogram channel vs. input signal (PMT signal [mV]) and the corresponding flow karyotype (a) linear, (b) logarithmic distribution. The measurements were made on an EPICS V flow cytometer. In both cases the chromosome preparation was the same as well as the operating parameters. The letters indicate the corresponding peaks. (Diploma-Thesis-permission given by the authors-figures B. Hagmann, 1993)

Figure VIII.4. Flow karyotype of chromosome suspensions of the Chinese hamster cell line CHL prepared according to chromosome preparation methods. (Diploma-Thesis permission given by the authors)
(a) Methanol-acetic acid method. (b) HEPES/MgSO$_4$ method. (c) HEPES modified method.

In all cases, the chromosomes have been stained with propidium iodide: (a), (b) 10 µM and (c): 75 µM for. In figures a)-b) the laser power was 1.1 W, in figure c) 500 mW. The ordinate shows the particles frequency, the abscissa the logarithm of the fluorescence intensity. In all cases, the number of registered particles was comparable. The letters indicate the peaks that can be resolved without further data and analysis (figures B. Hagman, 1993).

Figure VIII.6.a. Example of a two parameter flow karyotype (Hoechst 33258 fluorescence vs. chromomycin A3 fluorescence) of the cell line McCoy (human cell line with presumable mouse contamination). The chromosomes were prepared according to the modified Polyamine method. a) entire flow karotype; b) window on the smaller chromosomes. The numbers indicate peaks separated from the diagonal line corresponding to chromosomes with AT/GC ≠ 1. (Diploma-Thesis, permission given by the authors, figures C. Knoll, 1992)

Figure VIII.6.b. Two parameter flow karyotype (64 × 64 channels; Hoechst 33258 fluorescence vs. chromomycin A3 fluorescence) of human chromosomes. The fluorescence amplification was linear for both parameters. The contour lines show relative particle frequency (100, 62, 28, 13, 6, 3, 1, 0.5). The numbers identify the chromosome type. Due to the fluorescence gain used, the peaks of chromosomes 1–3 were excluded from display, which allows a better discrimination of the smaller chromosomes. (from Cremer et al., 1989b, reproduced with the permission of Berichte der Bunsen-Gesellschaft für Physikalische Chemie).

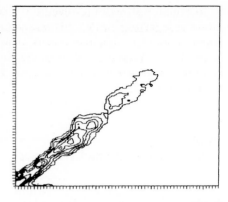

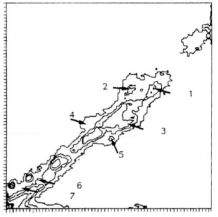

a

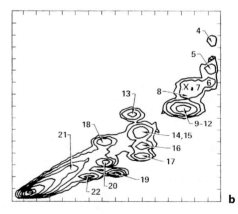

b

Another technique to improve the peak resolution can be applied in cases where the AT/GC ratio in different chromosome types is unequal 1 (e.g. human). In such cases two fluorescence parameters can be measured simultaneously (AT-specific fluorescence and GA-specific fluorescence). The particle frequency in two parameter histograms can be visualised by dot-frequency, altitude lines (Figure VIII.6), or pseudo three dimensional distributions.

VIII.4.2
Data evaluation of flow karyotypes

The peak pattern of a flow karyotype characterises the chromosome complement. In many cases flow karyotypes additionally contain a continuum of an unspecific background deriving from chromosome fragments and aggregates. For the quantitative evaluation of the peak pattern this background has to be subtracted by a suitable function. In Figure VIII.7 a program running on PC/AT of compatible systems is described – an example of flow karyotype of the cell line CHL (Dölle et al., 1991).

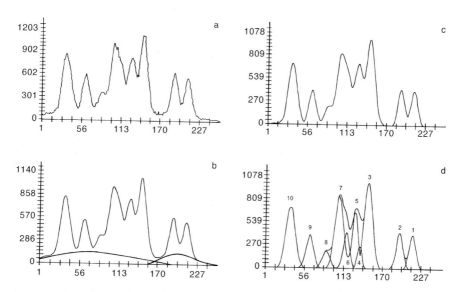

Figure VIII.7. Flow karyotype of the chromosomes of the Chinese hamster cell line CHL isolated according to the modified Hexyleneglycol method and stained with DAPI.
(a) original histogram (logarithmic scale on the abscissa);
(b) smoothed peaks with curves for background subtraction (X^2 function);
(c) resulting peak pattern;
(d) matching of Gaussian distribution curves (decomposition of the karyotype into 10 fit curves and the sum of the single peaks).

The table below shows the result of peak quantification (from Dölle et al., 1991, reproduced with the permission of Analytical Cellular Pathology).

The background subtraction depends on the application of X^2 functions:

$$X^2(x) = Ax^k \exp(-x/2)$$

x = fragment length.

k = free parameter;

A = norming factor.

For further quantification, gaussian curves can interactively be fitted to each peak of an experimentally 7 measured flow karyotype:

$$N(x) = \frac{A}{\sqrt{2\pi}\sigma} \exp\left(-\frac{(x-x_0)^2}{2\sigma^2}\right)$$

$N(x)$ = number of counts in channel $\times$;

x_0 = position of the maximum of the fit curve;

σ = standard deviation of the fit curve;

A = peak area.

Different peak values can be determined as for instance peak mean, standard deviation, absolute and relative peak area etc. For Figure VIII.7 these are summarised in the corresponding table. For the evaluation of two parameter flow karyotypes the procedure described above can be applied to the projections on the parameter axes. Also two dimensional Gaussian fit curves can be applied (Van den Engh et al., 1990).

VIII.4.3
Slit-scan measurements

In cases where chromosomes do not differ in their DNA content or DNA base pair composition further discrimination parameters have to be measured. One of these parameters may be the centromeric index

$$C_1 = \frac{\text{long arm fluorescence/length}}{\text{chromosome fluorescence/length}}$$

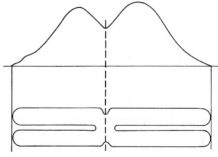

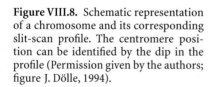

Figure VIII.8. Schematic representation of a chromosome and its corresponding slit-scan profile. The centromere position can be identified by the dip in the profile (Permission given by the authors; figure J. Dölle, 1994).

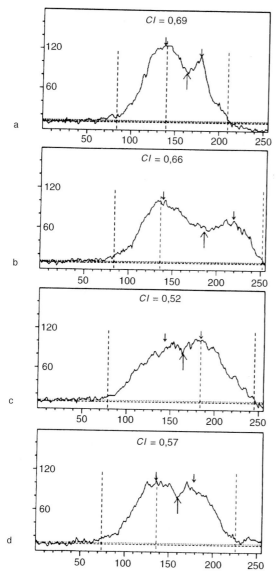

Figure VIII.9. Slit-scan profiles of isolated pig chromosomes prepared according to the TAcCaM method after staining with propidium iodide (relative fluorescence intensity vs. time of flight in units of 20 μseconds). a) and b) correspond to chromosome 1; c) and d) correspond to a marker chromosome t(6;15) of about the same DNA content and length. Both chromosomes can be discriminated by their centromeric index (CI). The horizontal dotted lines determine the level of fluorescence background. The vertical dashed lines show the calculated beginning, intensity maximum, and end of a profile. The big arrow indicates the automatically determined position of the centromere (from Hausmann et al., 1993, reproduced with the permission of Verlag der Zeitschrift für Naturforschung).

This can be obtained from slit-scan profiles of individual chromosomes (Figure VIII.8). In Figure VIII.9 an example for pig chromosomes is given. In this aberrant karyotype chromosome 1 was not discriminated from a marker chromosome t(6;15) by standard flow karyotype. However, both chromosomes differed significantly in their CIs [chromosome 1 CI = 0.65; t(6;15) CI = 0.52]. For the evaluation the profiles were registered by the Heidelberg slit-scan fluorometer and analysed automatically.

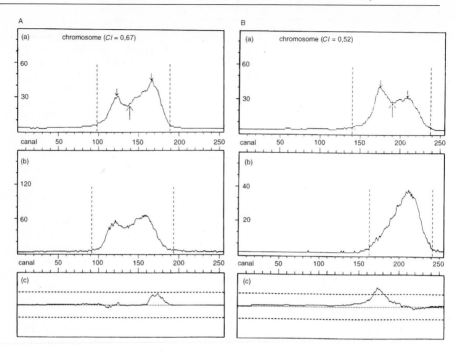

Figure VIII.10. Slit-scan profiles (relative fluorescence intensity vs. time of flight in units of *100 μsec*) of chromosomes of a Chinese hamster X human hybrid cell line after FISH in suspension of the human material (FITC labelling). A) normal human chromosome; B) interspecies translocation human Chinese hamster. From profiles a) (DAPI counterstaining) the chromosomes can be classified according to their centromeric index. From profiles b) (FITC label simultaneously registered) the human chromosomal material can be determined. Profile c) results from the difference of a) and b) after normalisation. With appropriate thresholds (dashed lines) an automatic chromosome classification was possible (from Hausmann et al., 1991, reproduced with the permission of Verlag der Zeitschrift für Naturforschung).

As a further slit-scan parameter the specific quantity of the fluorescent label, that is florescein on chromosomes after in situ hybridization in suspension can be measured simultaneously with the DNA counterstaining. Figure VIII.10 shows an example of a Chinese hamster × human cell line. In addition to DAPI staining the human material was labelled by FITC after FISH with gnomic DNA. Two profiles were registered simultaneously. Hamster chromosomes showed a 'considerable intensity on the DAPI profile. Human chromosomes showed two similar looking profiles (DAPI and FITC). Interspecies translocations showed a "normal" bimodal. DAPI profile but an "aberrant" fluorescein profile. Figure VIII.11 shows the result of the quantitative profile evaluation.

Another application of slit-scan flow fluorometry is the detection of rare events in a karyotype which may be under the background level in standard flow karyotypes. Such rare events could be dicentric chromosomes for instance after irradiation with ionising radiation. In these cases it should be

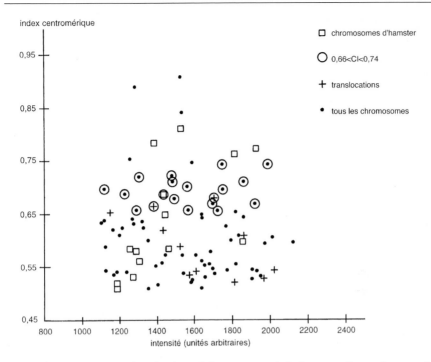

Figure VIII.11. Automated evaluation of the centromeric index according to integrated fluorescence intensity for a Chinese hamster x human hybrid cell line after FITC labelling of the human material by FISH (from Hausmann et al., 1991, reproduced with the permission of Verlag der Zeitschrift für Naturforschung).

helpful to sort out the rare events for further analysis (enrichment of rare chromosomes). For this purpose online profile analysis has to be performed. Figure VIII.12 shows two slit-scan profiles of a normal and dicentric chromosome. In both cases a second profile with several thresholds is shown. This is the second derivative of the intensity profile. This derivative has a characteristic pattern for normal chromosomes, dicentric chromosomes, and debris. Thus it can be used for profile classification.

In Figure VIII.13 the application of this fast online algorithm to pig chromosomes is shown. The slit-scan results were compared to microscopic measurements obtained from the same cell line. Down to the resolution limit of the slit-scan flow fluorometer both results were in good correlation.

VIII.4.4
Perspective

Although flow cytometers are commercially available for clinical routine e.g. in immunology or haematology, flow cytometry of chromosomes is still far from daily routine. It requires great experience in the preparation of chromosome suspensions and needs sophisticated optics, electronics, and data analysis. However, flow karyotyping together with slit-scanning is a versatile

Figure VIII.12. Slit scan profiles of a meta-centric (a) and dicentric (b) chromosome. For evaluation an online data analysis program was used. It is based on a threshold procedure on the second derivative (profile below). The vertical lines in the chromosome profile indicate the points of inflection (beginning and end) and the maximum intensity. This classification procedure appeared to be very stable against variations in measuring conditions. (Permission given by the authors, figure J. Dölle, 1994)

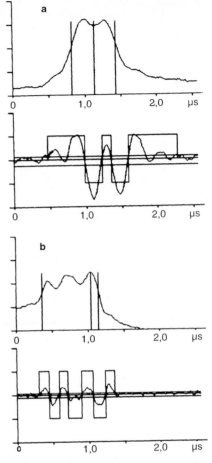

tool for a fast analysis and for sorting of chromosomes in mammalian cyto-genetics and biological dosimetry (Hausmann et al., 1995; Hausmann et al., 1996a). The development of new preparation protocols and staining tech-niques (e.g. chromosome painting in suspension) will offer new perspectives in the application of flow cytometry to further species.

VIII.4.5
Identification of the sorted chromosomes by GTG banding

P. POPESCU

Principle
Chromosomes obtained after the flow sorting are free in suspension by thou-sands in the sorting buffer, in contrast to chromosome classical preparation which preserves the cytoplasmic membrane of the cell spreads. The devel-oping of the banding patterns on these kinds of preparations is more

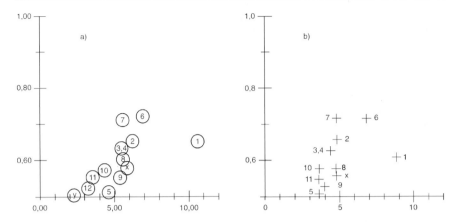

Figure VIII.13. Comparison of microscopic measurements a) and slit-scan evaluation of profiles b) using the online evaluation program based on the threshold procedure on the second derivative. The graphs shows centromeric index vs. chromosome length. Due to the evaluation procedure used for the chromosome length from slit-scan profiles (beginning and end = points of inflection) the calculated values are always lower than measured in microscopy. Both data sets show a good correlation down to the resolution limit of the slit-scan fluorometer. (Permission given by the authors, figures J. Dölle, 1994)

difficult, as the chromosome structure is often not very well preserved. The sorted chromosomes banding is thus necessary to verify the sorting purity or to identify the chromosome of each peak. Banding pattern developing can be useful (Figures VIII.14 and VIII.15), e.g. to differentiate an abnormal and a normal chromosome having similar sizes and centromeric indexes (Hausmann et al., 1993), also in slit scan cytometry which allows calculation of chromosomes length and centromeric index.

Several banding methods have been applied to sorted chromosomes: QFQ banding, DAPI staining, but the best and most reproducible technique is the GTG banding.

Protocol
The sorted chromosome suspension obtained at the flowing cell stream is spread on clean, dried slides and stained with Giemsa. After chromosome localisation by low magnification microscope observation (25× or 40×), the slides are faded by washing with absolute ethanol and air dried. The GTG banding is performed by the treatment of the slides in a 0.25% trypsin solution (appendix S5) and restained with Giemsa for 10 minutes (according to paragraph I.5.1).

Remarks
The use of buffer TAcCaM (according to paragraph VIII.3.2), which preserves the chromosome structure is highly recommended to obtain a qualitative GTG banding of the sorted chromosomes (Hausmann et al., 1993).

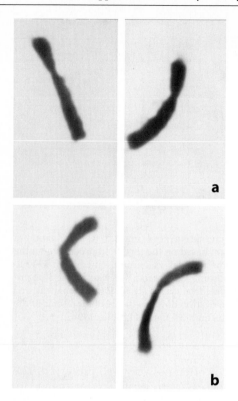

Figure VIII.14. Pig chromosomes sorted: (a) chromosome 1 normal and (b) translocation t(6;15) (Hausmann et al., 1993, reproduced with the permission of Verlag der Zeitschrift für Naturforschung).

VIII.5
In situ hybridisation to chromosomes in suspension

D. CELEDA

Principle
The hybridisation of specific DNA probes to isolated metaphase chromosomes in suspension however, may offer a new approach to chromosome analysis and chromosome separation. First investigations have been carried out on chromosomes obtained from a Chinese hamster × human hybrid cell line using human genomic DNA as the probe (Dudin et al., 1987, 1988; Hausmann et al., 1991).

The techniques for FISH in suspension on isolated chromosomes represent a modification of FISH usually used for metaphase chromosomes and interphase nuclei fixed on slides (Figure VIII.16). Formamide (and to some extent dextransulfate) is considered as an established component of this method. However, a certain number of washing steps are necessary after hybridisation. In particular the washing steps concerning FISH in suspension are exclusively based on centrifugations. This procedure lead to a considerable reduction of the final amount of chromosomal material, approx. up to

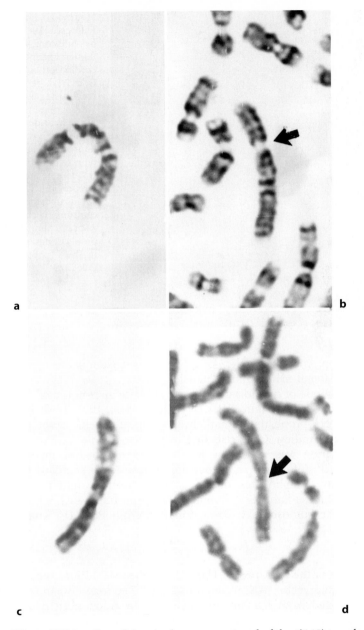

Figure VIII.15. GTG banding of the pig chromosome 1 and of the t(6;15) translocation: (a) and (c) on the sorted chromosomes and (b) and (d) on the chromosome spreads on the slides prepared using classical methods (photographs P. Popescu).

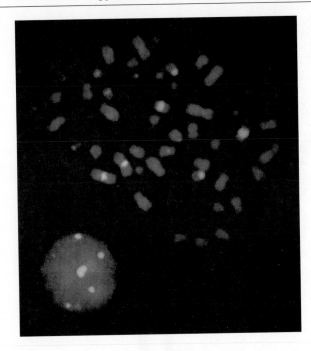

Figure VIII.16.
Fluorescent in situ hybridisation on fixed human metaphase chromosomes obtained from fresh peripheral blood using centromeric DNA probes. Visualisation of the hybridisation by Anti-digoxigenin~ fluorescein Fab fragments, representing the pUC 1.77 DNA-probe and Streptavidin Texas Red conjugated, representing a centromeric chromosome 8 specific human DNA probe. Counter-staining by DAPI (photographs D. Celeda).

30% per step. Another problem of FISH in suspension is the aggregation of the chromosomal material in suspension.

The technique of hybridisation described here does not need formamide and dextransulfate – Figure VIII.17 (Celeda et al., 1992, 1994; Haar et al., 1994). As DNA-probe, the human chromosome 1 specific probe (pUC 1.77: Cooke and Hindley, 1979) labelled with digoxigenin and/or avidin by nick translation was used. The labelled DNA probe was hybridised on metaphase chromosomes in suspension, prepared by standard techniques from human lymphocytes obtained from fresh peripheral blood and from the Chinese hamster × human hybrid cell line A_3 IG 5-1, containing the human chromosome 1 (kind donation from Prof. Dr. Khan, Leiden, Netherlands).

Protocol

■ **Labelling of the DNA-probe.** The DNA-probe pUC 1.77 designates a clone of the plasmid vector pUC 9 containing a 1. 77 kb long human fragment as insert, isolated from the human satellite DNA fraction II/III. The insert mainly represents a tandem organised repetitive sequence in the region q 12 of the human chromosome 1 (Cooke and Hindley, 1979). The DNA -probe pUC 1.77 was labelled with digoxigenin and biotin by nick translation according to standard techniques (digoxigenin for fluorescent detection; biotin for magnetic separation).

■ **Preparation of the chromosome in suspension.** The preparation of the chromosomes from human lymphocytes obtained from fresh peripheral

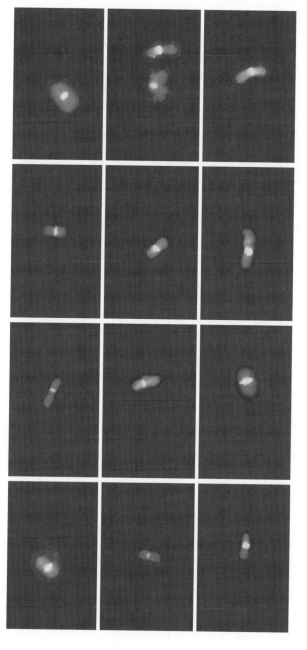

Figure VIII.17. Isolated human chromosomes after FISH (fluorescent in situ hybridisation) in suspension according to the technique described in paragraph VIII.5. Labelling of the pUC 1.77 DNA probe was done by using DIG(digoxigenin)-11-dUTP and detected with Anti-digoxigenin-fluorescein Fab fragments. The hybridisation signals on the chromosomes are located in the centromeric region, corresponding to the main hybridisation site of the pUC 1.77 DNA probe (1q12). Counterstaining with propidiumiodide and DAPI (photographs D. Celeda).

blood was performed according to standard techniques with the following modifications for FISH in suspension: the chromosome preparation is stored in methanol 3 vol.-acetic acid-1 vol. at −20 °C. Best results are achieved when the suspended chromosomes are stored no more than 10 days prior to hybridisation.

■ *In situ* **hybridisation in suspension.** 1 volume of the chromosome suspension containing approx. 7×10^4 chromosomes is transferred in a 500 µl Eppendorf tube. If necessary, the remaining metaphase spread is sonicated by incubating the Eppendorf tube in a sonicator with a frequency of 41 kHz and an initial power of 30 W, for 10 sec. The suspension is centrifuged for 10 min at 350 g, the supernatant is discarded and the chromosome pellet air dried. 100 µl of the hybridisation mix containing 20 µl buffer solution, [TrisHCl, pH 7, 150 Mm (appendix SS14); MgCl₂, 15 mM (appendix SS9); Tween 20, 0.025%; Nonidet p-40, 0.025%], 10 µl 20× SSC buffer (appendix B4), 200 ng labelled DNA-probe and deionized distilled water, to achieve a final volume of 100 µl, are pipetted to the air dried chromosome pellet in the 500 µl Eppendorf tube and resuspended carefully. This hybridisation mixture is overlaid with 40 µl paraffin oil and denatured in a water bath at 90 °C for 5 min. The hybridisation is renatured in a water bath having an initial temperature of 72 °C and allowed to cool down to 40 °C overnight.

■ **Post hybridisation preparations wash steps.** The suspension is centrifuged after the hybridisation at 350 g for 10 min. and the supernatant is discarded, always being aware to discard only up to approx. vol. of 2 mm above the pellet. The pellet is resuspended to 100 µl of a wash solution [4× SSC buffer pH 7 (appendix B4); 0.02% Tween 20] and centrifuged for 10 minutes at 350 g. Blocking: the remaining aliquot is resuspended in 100 µl of a solution (4× SSC buffer pH 7; 0.2% Tween 20; 3% bovine serum albumin) and incubated at 37 °C for 30 min. The solution is centrifuged and the supernatant discarded.

■ **Detection by immunofluorescence.** The suspension is centrifuged for 10 minutes at 350 g and resuspended in 200 µl Omnibuff solution (WAK Chemie Medical GmbH, JenaBioTech, Germany) or (4× SSC buffer – 0.2% Tween 20) containing the antibody antidigoxigenin coupled with an appropriate fluorescent dye, at the concentration proposed by the supplier.

The 200 µl is transferred to the aliquot with the hybridised and blocked chromosomes and incubated for 60 min at 37 °C. The mixture is centrifuged again for 10 minutes at 350 g, the supernatant discarded and the remaining aliquot is resuspended in 100 µl of solution (4× SSC buffer; 0.02% Tween 20).

For evaluation of the hybridisation, 5 µl of the final suspension containing the hybridised chromosomes is pipetted on a microscopic slide and counterstained with 5 µl DAPI at 1.75 µg/ml (appendix S2) and propidium iodide at 10 µg/ml (appendix S12). The slide is rinsed and covered with a coverslip ($20 \times 40\ mm^2$), then observed under fluorescence microscope.

■ **Sorting of hybridised chromosomes with magnetic particles.** Sorting of chromosomes was achieved using magnetic particles, Dynabeads^R M-450 (Dynal AS, Oslo/Norway). They consist of polystyrene with hydroxyl groups and an inner Fe_3O_4 core (Ugelstadt et al., 1983). The beads are coated with streptavidin and attached to the biotynilated DNA probe (pUC 1.77) which has been prior hybridised to the human chromosome 1 in suspension. This sorting allows the separation of the human chromosome 1 from other chromosomes.

The streptavidine free magnetic beads are, at first activated. The beads (4×10^8 beads = 30 mg/ml) are washed for 5 min each at room temperature in a solution consisting of the following components:

- 7 vol. distilled water – 3 vol. acetone: 1 time
- 6 vol. distilled water – 4 vol. acetone: 1 time
- 2 vol. distilled water – 8 vol. acetone: 1 time
- acetone: 3 times

After each wash step the beads are collected with a permanent magnet according to the protocol indicated in the Figure VIII.18 (Dynal-MPC 1, Dynal AS, Oslo Norway) and the remaining suspension is discarded. The beads are resuspended in pure acetone for storage. 1 ml of the beads, resuspended in acetone is mixed with 0.5 µl of 1.5 mM pyridine and 0.5 ml of 1.5 mM P-toluensulfonyl-4-chlorid (286 µg/ml) The reagents react with the hydroxyl group of the surface of the beads and form a-toluensulfonyl-4-chlorid ester. For the formation of the ester, the beads are incubated for 20 hours at room temperature under permanent rotation. The activated beads are collected with a permanent magnet, washed 3 times in fresh acetone (10 ml) for 5 min each at room temperature and washed according to the above protocol but reversing the steps, that is beginning with the end and finishing with the first step. Then the beads are resuspended in distilled water, incubated for 5 min, collected with a permanent magnet and resuspended in 10 ml of 1 mM HCL-solution. The beads can now be stored at 4 °C for up to 12 months.

The activated beads are then treated with streptavidin. The streptavidin is dissolved in a concentration of 150 µg/ml in 0.2 M borate buffer (disodium tetraborate, H_2O free, 4.02 g/100 ml). The beads stored in 1 mM HCL solution are collected with a permanent magnet and washed twice in 10 ml of distilled water for 5 min each. 1 ml or 30 mg of beads are added to 1 ml streptavidin solution at a concentration of 150 µg/ml. The mixture is incubated under permanent rotation for 24 hours in order to achieve the formation of a covalent link between the streptavidin molecules and the activated beads. The beads are collected with a permanent magnet and resuspended in KCL-HEPES-MgSO₄ buffer (appendix F14) containing 0.02 g bovine serum albumin in order to achieve the desired final concentration (preferably 30 mg/ml). The beads are now coated with streptavidin and can be stored for up to 6 months.

The linkage of streptavidin coated magnetic beads and the biotinilated probe hybridised on chromosomes in suspension is achieved by the mixture of the two suspensions in a way that the bead/hybridised chromosome ratio is = 2/1. This mixture is then incubated for 2 hours minimum (not more than one night) at room temperature. The chromosomes, attached to beads are collected with a permanent magnet and resuspended in a suitable buffer. For observation under a microscope, pipette 5–10 µl of the collected chromosomes on a microscopic slide and counterstain with a suitable DNA dye, e.g. DAPI (Figure VIII.19).

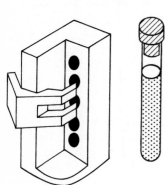

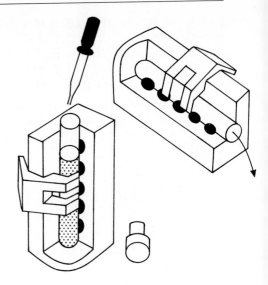

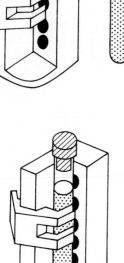

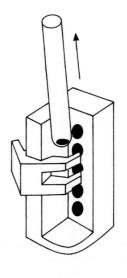

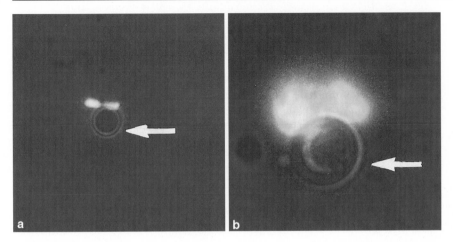

Figure VIII.19. Human chromosome after magnetic separation linked to a magnetic microsphere (0.5 µm diameter) – indicated by an arrow, and FISH (in situ hybridisation) in suspension with biotinylated pUC 1.77 DNA probe. Counterstaining with DAPI.
(a) Chromosome from a normal human line.
(b) Human chromosome 1 separated from a chromosome suspension obtained from the Chinese hamster X human cell hybrid line, A$_3$IG 5-1 (photographs D. Celeda).

Figure VIII.18. Instruction manual of the permanent magnet Dynal MPC-1 (Dynal AS, Oslo, Norway).

1 The test tube containing cells and Dynabeads M-450 coated with specific antibodies is mixed and incubated according to the experimental protocol.
2 Place the test tube in the DYNAL MPC1 and ensure that the clamp holds the tube.
3 Let the test tube stay in the DYNAL MPC1 for 1–2 minutes. During this time the cells rosetted with immunomagnetic beads (Dynabeads) will be attracted to the wall of the test tube by the magnetic field.
   The magnetic isolation may be accelerated by an occasional gentle shaking of the DYNAL MPC1 while the test tube is held in the MPC1.
4 Remove the supernatant by either pipetting or pouring it out of the test tube while this is still held in the DYNAL MPC1. The immuno-magnetic rosetted cells are kept on the wall of the test tube by the magnetic field.
5 Washing of rosetted cells.
A After discarding the supernatant in step 4, remove the test tube from the DYNAL MPC1. Add wash solution and resuspend the isolated cells rosetted with immunomagnetic beads (Dynabeads M-450) coated with antibodies. Place the tube in the Dynal MPC1 for a minimum of 30 seconds between each washing cycle or until the wash fluid is clear.
Repeat the washing step 4–5 times.
B Discard wash solution.
C Concentrate the isolated cells as a pellet at the bottom of the test tube by slowly removing the tube from the DYNAL MPC1. The cells will slowly move towards the bottom of the tube as this is moved out of the holder.
D The cell pellet can now be used for further studies.

Remarks

The purity degree of the suspension of the sorted chromosomes (sorting of magnetic particles) can be increased. The suspension is decanted for 15 minutes before the contact with the beads. This procedure allow to eliminate the accretion and the interphasic nucleus from the supernatant. The purity degree can be compared with those obtained by classical fluorometry. This technique is very cheap and does not require a heavy instrumentation.

The chromosomes sorted by this method are used for the specific DNA banks.

Appendix

S Solutions for chromosome staining and banding techniques

S1 0.3 N Ba(OH)$_2$ solution

A solution of Ba(OH)$_2$ (9 g/100 ml H$_2$O) is prepared by heating, which corresponds to a saturated solution at room temperature. Barium hydroxide Ba(OH)$_2$ carbonates spontaneously in contact with air, producing a white film at the surface of the solution.

S2 DAPI (4′-6-Di-Amidino-2-Phenyl-Indole) solution

stock solution (1 mg/ml): 1 mg of DAPI in 1 ml of distilled H$_2$O.
This solution can be stored in 50 or 100 µl aliquots at −20 °C and in the dark for several months.
working solution (0.4 µg/ml): 10 µl of the stock solution are diluted in 25 ml of MacIlvaine's buffer pH 7 (appendix B6).

S3 DA (distamycin A-HCl) solution

0.2 mg/ml solution:
- DA: 10 mg
- MacIlvaine's buffer pH 7 (appendix B6): 50 ml
This solution should be prepared freshly before use.

S4 50% silver nitrate solution

5 g of silver nitrate (AgNO$_3$) are dissolved in 10 ml of H$_2$O containing 2 drops of formaldehyde. The solution must be stored in the dark.

S5 0.25% or 0.1% trypsin solution

The following solution is prepared extemporaneously:
- 250 mg or 100 mg of trypsin (1:250 Difco)
- PBS⁻ buffer (appendix B1) to 100 ml.

S6 Hoechst 33258 1× solution

100× stock solution (1 mg/ml): 100 mg of Hoechst (Sigma B2883) in 100 ml of
H_2O. Store in 1 ml aliquots at −20 °C.
1× working solution (10 µg/ml): dilute 1 ml of the 100× solution in 99 ml H_2O.

S7 Giemsa solution

The solution is prepared extemporaneously:
– Sorensen's buffer pH 6.8 (appendix B3): 3 ml
– Giemsa rapid stain or Giemsa R (in solution, R.A.L. Prolabo, Paris):
  3 ml
– distilled water: 94 ml.

S8 acridine orange solution

stock solution (1 mg/ml): 100 mg of acridine orange are dissolved in 100 ml of
distilled H_2O. This solution can be stored at 4 °C in the dark for several
months.

S9 1× quinacrine mustard solution

100× stock solution (2.5 mg/ml): 25 mg of quinacrine mustard are dissolved
in 10 ml of distilled H_2O. Store in 1 ml aliquots at 4 °C in the dark (up to 6
months).
1× solution (50 µg/ml): dilute 1 ml of the 100× solution in 49 ml of distilled
H_2O.

S10 PPD11 solution

9 ml of bidistilled glycerol are mixed with 1 ml of phosphate buffer pH 7.2–7.4
(appendix B10), 10 mg of PPD (p-phenylenediamine, Sigma P6001) are added
and dissolved by heating at 37 °C. After cooling, the pH is adjusted to 11 with
about 10 drops of 5 M NaOH solution (appendix SS12).
Store at −20 °C in the dark.

S11 PPD7 and PPD8 solutions

9 ml of bidistilled glycerol are mixed with 1 ml of phosphate buffer pH 7.2–7.4
(appendix B10), 10 mg of PPD (p-phenylenediamine, Sigma P6001) are added
and dissolved by heating at 37 °C. After cooling, the pH is adjusted to 7 for
PPD7 and to 8 for PPD8.
Store at −20 °C in the dark.

S12 propidium iodide solution

1000× stock solution (1 mg/ml): 10 mg of propidium iodide in 10 ml of H_2O.
Store in 100 µl aliquots at −20 °C in the dark.
1× working solution (1 µg/ml): 1 µl of the 1000× solution is diluted in 1 ml of
phosphate buffer (appendix B10) or of PBT solution (appendix H3) and kept
on ice in the dark.

S13 50% AgNO₃ solution

- 50% AgNO₃: 2.5 g
- distilled H₂O: 5 ml

This solution should be prepared at least two days before use. It can be stored in the dark for 3 to 4 weeks. If crystals appear, the solution should be renewed.

S14 ammoniacal silver nitrate solution

- AgNO₃: 2 g
- distilled H₂O: 2.5 ml
- 28% NH₄OH: 2.5 ml

This solution can be stored in the dark for 3 to 4 weeks. If crystals appear, the solution should be renewed.

S15 3% formalin solution

- 37% formaldehyde solution: 3 ml
- distilled H₂O: 97 ml

The pH is adjusted to 7 by addition of sodium acetate crystals and then just before use it is adjusted to 4.5 with a few drops of formic acid.
Store the solution in the dark.

S16 chromomycin A3 solution

stock solution (0.5 mg/ml):
- chromomycin A3: 5 mg
- MacIlvaine's buffer pH 7 (appendix B6): 5 ml
- 50 mM MgCl₂ solution (1/20 dilution of a 1 M solution, appendix S9): 0.1 ml
- distilled H₂O: 4.9 ml

Chromomycine A3 is dissolved slowly without stirring in this buffer solution at 4 °C for one night. This solution can be stored in the dark for several months.

M Miscellaneous

M1 75 mM KCl solution

1× solution: 5.6 g of KCl in 1 litre of distilled H₂O. The 1× solution is diluted 1/2, 1/10 etc . . . according to the desired concentration.
The solution can be stored at room temperature for 2 to 3 weeks.

M2 acetic violet stain

Prepare the following mixture:
- crystal violet (Sigma C3886): 0.02 g
- acetic acid: 1 ml
- H₂O to 100 ml

10 µl of the cell suspension are mixed with 190 µl of the staining solution. In this case, the red blood cells are lysed.

M3 trypan blue stain

Mix:
– 100 µl of culture medium
– 50 µl of trypan blue solution (Sigma T8154)
– 50 µl of cell suspension
The mixture is placed in a Malassez counting chamber, live cells are not stained while dead cells are stained blue.

M4 genomic DNA preparation from whole blood

Blood (30 ml) is collected in tubes containing disodium EDTA and transferred to 50 ml centrifugation tubes. 20 ml of a solution of NaCl 10 mM – EDTA 10 mM pH 7.5 are added and incubation is carried out at 4 °C for one hour under horizontal stirring to lyse the red blood cells. The mixture is then centrifuged at 13 200 g for 15 minutes. This operation is carried out twice to obtain a white blood cell pellet separated from the red blood cell debris. The pellet is dispersed by vortexing and resuspended in the following mixture:
– 14 ml of 6 M guanidine chloride (to denature the proteins)
– 1 ml of 7.5 M ammonium acetate
– 1 ml of 20% NLS (detergent)
– 4 mg of proteinase K (protein digestion)
The suspension is mixed and incubated at 55 °C for one night under horizontal stirring to lyse white blood cells and to free the nuclear DNA. Three volumes of cold ethanol are added which precipitates the DNA as a fibrous mass which is collected by spooling on a Pasteur pipette. The collected DNA is washed twice in 70% cold ethanol, dried and finally resuspended in 2 ml 1× TE buffer (appendix B5) and the suspension is stirred gently at 4 °C for at least one night.
DNA is sonicated into fragments ranging in size from 300 to 500 base pairs, the average size of the fragments is controled by electrophoresis in a 1% agarose gel. After phenol and chloroform/isoamylalcohol extraction, the DNA is precipitated and dried, and resuspended in 1× TE buffer (appendix B5). The quantity obtained is estimated by spectrometry. The DNA is diluted to a concentration of 5 µg/µl and stored in 100 µl aliquots at −20 °C.

M5 slide washing for chromosome spread preparations

Slides are immersed in 1/3 diluted sulphuric acid for one night, rinsed in running water, and then rinsed one by one in distilled water. They are stored in distilled water at 4 °C.
If prewashed Esco slides are used, washing is not necessary.

M6 slide degreasing for synaptonemal complex spread preparations

Slides should always be handled with tweezers at the edges and always at the same place to avoid any damage.
Slides are immersed in 10% hydrochloric acid for 20 minutes then rinsed twice with bidistilled water (slides are immersed and the water discarded

twice). They are then soaked in 1% Teepol detergent for 20 minutes, rinsed three times in distilled water, and air dried vertically for half a day covered with filter paper to protect them from dust.

M7 deionisation of formamide

1 litre of formamide is stirred for about 4 hours with 50 g of Dowex MB-X8B 20–50 mesh ion exchange resins (from Touzart and Matignon). After filtration on Whatman paper, the pH of the formamide is controlled by preparing a 2× SSC solution pH 7 containing 50% of formamide. The formamide is stored in 35 ml aliquots at −20 °C in the dark.

M8 0.5% optilux plastic solution and slide preparation

An optilux Falcon Petri dish is broken into small fragments by wrapping it in filter paper and crushing it with hands protected by clean latex gloves. 0.5 g of this plastic is dissolved in 100 ml of chloroform in an air tight jar, to avoid chloroform evaporation.
Two days before the preparation of synaptonemal complexes, the slides are coated with optilux plastic by dipping them slowly one by one in a borel tube containing 0.5% optilux solution and holding them vertically with a tweezer to let the bubbles come to the surface. Each slide is then taken out very slowly avoiding any waves to produce a uniform film. Slides dry instantaneously. They are stored protected from dust. Before use, each slide should be examined to eliminate any presenting defects in the plastic film and a test of film release should be carried out on one slide.

M9 fixative (preparation of synaptonemal complexes)

13.6 g of sucrose are dissolved in 280 ml of distilled H_2O, 16 g of paraformaldehyde are added and the solution is completed to a final volume of 400 ml with H_2O. This solution is filtered through ordinary filter paper, 24 drops of 1 M NaOH (appendix SS12) are added and the mixture is heated slowly to 60 °C with gentle stirring until it becomes clear. After cooling, the pH is adjusted from 9.5 to 8.5 with 0.1 M boric acid (appendix SS1). The solution may be stored for several days at 4 °C and should be filtered 30 minutes before use.

M10 0.4% photoflo solution

4% photoflo stock solution (ethylene-glycol, Kodak Photoflo 600 solution).
0.4% working solution: dilute 4 ml of the stock solution in 96 ml of distilled H_2O. The pH is adjusted from 6–7 to 8.5 with borate buffer (appendix B7). The solution is filtered just before use.

M11 hyaluronidase solution

stock solution (10 mg/ml)
– hyaluronidase (Sigma H 3506): 500 mg

- PBS⁻ (appendix B1): 50 ml

Filter through a 0.22 µm membrane and store in 0.5 ml aliquots at −20 °C.
working solution (1 mg/ml): dilute 0.5 ml of the stock solution in 4.5 ml of
PBS⁻ solution containing 20% FCS (foetal calf serum).

M12 0.85% KCl hypotonic solution

- KCl: 0.85 g
- distilled H_2O: 100 ml

Filter through a 0.22 µm membrane and store at room temperature.

M13 0.05% trypsin – EDTA 0.02% solution

- trypsin: 0.5 g
- EDTA: 0.2 g
- PBS⁻ (appendix B1) to 100 ml

Filter through a 0.22 µm membrane under sterile conditions and store at
−20 °C.

This solution is available commercially (Gibco 45300 019).

L Solutions for lampbrush chromosomes

L1 rinsing solution

- 82.5 mM NaCl: 8.25 ml of 1 M NaCl (appendix SS11)
- 2.5 mM KCl: 2.5 ml of 0.1 M KCl (appendix SS8)
- 0.1 mM $CaCl_2$: 0.1 ml of 0.1 M $CaCl_2$ (appendix SS7)
- 1 mM $MgCl_2$: 1 ml of 0.1 M $MgCl_2$ (appendix SS9)
- 1 mM Na_2HPO_4: 1 ml of 0.1 M Na_2HPO_4 (appendix SS13)
- 0.125 mM polyvinylpyrrolidone PVP 40000: 1.25 ml of 10 mM PVP 40000
  (10 mM = 0.4 g/ml)
- 5 mM HEPES: 50 µl of 1 M HEPES (appendix SS2)
- distilled H_2O to 100 ml.

L2 1 mg/ml collagenase

- collagenase type I: 10 mg of collagenase type I (Sigma C0130)
- L1 solution (appendix L1): 10 ml.

L3 1 mM EDTA solution

- 82.5 mM NaCl: 8.25 ml of 1 M NaCl (appendix SS11)
- 2.5 mM KCl: 2.5 ml of 0.1 M KCl (appendix SS8)
- 1 mM Na_2HPO_4: 1 ml of 0.1 M Na_2HPO_4 (appendix SS13)
- 0.125 mM polyvinylpyrrolidone PVP 40000: 1.25 ml of 10 mM PVP 40000
  (10 mM = 0.4 g/ml)
- 5 mM HEPES: 0.5 ml of 1 M HEPES (appendix SS2)
- 1 mM $Na_2EDTA.2H_2O$: 1 ml of 0.1 M $Na_2EDTA.2H_2O$ (appendix SS5)
- distilled H_2O to 100 ml.

L4 saline solution

- 10 mM Tris: 1 ml of 1 M Tris-HCl pH 7.2 (appendix SS14)
- 75 mM KCl: 7.5 ml of 1 M KCl (appendix SS8)
- 25 mM NaCl: 2.5 ml of 1 M NaCl (appendix SS11)
- distilled H_2O to 100 ml.

L5 saline solution

- 25 mM NaCl: 2.5 ml of 1 M NaCl (appendix SS11)
- 0.05 mM $CaCl_2$: 50 µl of 0.1 M $CaCl_2$ (appendix SS7)
- 10 mM Tris: 1 ml of 1 M Tris-HCl pH 7.2 (appendix SS14)
- 75 mM KCl: 7.5 ml of 1 M KCl (appendix SS8)
- 1 mM $MgCl_2$: 1 ml of 0.1 M $MgCl_2$ (appendix SS9)
- distilled H_2O to 100 ml.

F Solutions for flow and slit-scan cytometry

F1 Tris-NaCl-MgCl$_2$ hypotonic solution

- 10 mM Tris: 10 ml of 0.1 M Tris-HCl pH 7.2 (appendix SS14)
- 10 mM NaCl: 10 ml of 0.1 M NaCl (appendix SS11)
- 5 mM $MgCl_2$: 5 ml of 0.1 M $MgCl_2$ (appendix SS9)
- final pH adjusted to 7.5 with 1 N HCl
- distilled H_2O to 100 ml.

F2 hexanediol solution

- 750 mM 1,6-hexanediol: 8.84 g
- 25 mM Tris: 25 ml of 0.1 M Tris-CH_3COOH pH 7.5 (appendix SS14)
- 5 mM $MgCl_2$: 5 ml of 0.1 M $MgCl_2$ (appendix SS9)
- 5 mM $CaCl_2$: 5 ml of 0.1 M $CaCl_2.2H_2O$ (appendix SS7)
- final pH adjusted to 3.2 with 1 M CH_3COOH
- distilled H_2O to 100 ml.

F3 TAcCaM buffer

- 25 mM Tris: 25 ml of 0.1 M Tris-CH_3COOH pH 7.5 (appendix SS14)
- 5 mM $MgCl_2$: 5 ml of 0.1 M $MgCl_2$ (appendix SS9)
- 5 mM $CaCl_2$: 5 ml of 0.1 M $CaCl_2.2H_2O$ (appendix SS7)
- final pH adjusted to 3.2 with 1 M CH_3COOH
- distilled H_2O to 100 ml.

F4 Tris-MgCl$_2$-ETB-MA-RNase hypotonic solution

- 10 mM Tris: 1 ml of 0.1 M Tris-CH_3COOH pH 7.5 (appendix SS14)
- 15 mM $MgCl_2$: 1.5 ml of 0.1 M $MgCl_2$ (appendix SS9)
- 25 µM ethidium bromide ETB: 25 µl of 10 mM ETB (appendix SS3)
- 25 µM mithramycin A: 250 µl of 1 mM mithramycin A (1 mM = 1 mg/ 0.92 ml)

- 25 mg/ml RNase: 250 mg of RNase
- distilled H_2O to 10 ml

Store in 1 ml aliquots at $-20\,°C$.

F5 Tris-MgCl$_2$-NaCl-ETB-MA solution

- 10 mM Tris: 1 ml of 0.1 M Tris-CH$_3$COOH pH 7.5 (appendix SS14)
- 15 mM MgCl$_2$: 1.5 ml of 0.1 M MgCl$_2$ (appendix SS9)
- 20 mM NaCl: 2 ml of 0.1 M NaCl (appendix SS11)
- 25 µM ethidium bromide ETB: 25 µl of 10 mM ETB (appendix SS3)
- 25 µM mithramycin A: 250 µl of 1 mM mithramycin A (1 mM = 1 mg/ 0.92 ml)
- distilled H_2O to 10 ml

Store in 1 ml aliquots at $-20\,°C$.

F6 P1 solution

- 15 mM Tris: 1.5 ml of 0.1 M Tris-CH$_3$COOH pH 7.2 (appendix SS14)
- 10 mM KCl: 1 ml of 0.1 M KCl (appendix SS8)
- 20 mM NaCl: 2 ml of 0.1 M NaCl (appendix SS11)
- 14 mM 2-mercaptoethanol: 10 µl of 2-mercaptoethanol d = 1.114 g/ml
- 2 mM EDTA: 200 µl of 0.1 M EDTA (anhydrous free acid) (appendix SS4)
- 0.5 mM EGTA: 50 µl of 0.1 M EGTA (free acid) (appendix SS6)
- distilled H_2O to 10 ml.

F7 P2 solution

- 15 mM Tris: 1.5 ml of 0.1 M Tris-CH$_3$COOH pH 7.2 (appendix SS14)
- 80 mM KCl: 0.8 ml of 1 M KCl (appendix SS8)
- 20 mM NaCl: 2 ml of 0.1 M NaCl (appendix SS11)
- 14 mM 2-mercaptoethanol: 10 µl of 2-mercaptoethanol d = 1.114 g/ml
- 2 mM EDTA: 200 µl of 0.1 M EDTA (free acid anhydrous) (appendix SS4)
- 0.5 mM EGTA: 50 µl of 0.1 M EGTA (free acid) (appendix SS6)
- 0.5 mM spermidine: 50 µl of 0.1 M spermidine (free base) (0.1 M = 14.5 mg/ml)
- 0.2 mM spermine: 20 µl of 0.1 M spermine (free base) (0.1 M = 20.2 mg/ml)
- 0.1% digitonin: 20 µl of a 50% digitonin solution
- distilled H_2O to 10 ml.

F8 PM1 solution

- 15 mM Tris: 1.5 ml of 0.1 M Tris-HCl pH 7.2 (appendix SS14)
- 50 mM KCl: 0.5 ml of 1 M KCl (appendix SS8)
- 20 mM NaCl: 2 ml of 0.1 M NaCl (appendix SS11)
- 0.5 mM spermidine: 50 µl of 0.1 M spermidine (free base) (0.1 M = 14.5 mg/ml)
- 0.2 mM spermine: 20 µl of 0.1 M spermine (free base) (0.1 M = 20.2 mg/ml)
- distilled H_2O to 10 ml.

F9 PM2 solution

- PM1 solution (appendix F8): 10 ml
- digitonin: 10 mg.

F10 HEPES-MgSO₄-KCl-DTT-RNase-PI hypotonic solution

- 50 mM KCl: 0.5 ml of 1 M KCl (appendix SS8)
- 5 mM HEPES: 50 µl of 1 M HEPES (free acid) (appendix SS2)
- 10 mM $MgSO_4$: 1 ml of 0.1 M $MgSO_4$ (appendix SS10)
- 3 mM dithiothreitol DTT: 30 µl of 1 M DTT (1 M = 154.2 mg/ml)
- 25 µg/ml RNase: 25 µl of 10 mg/ml RNase solution
- 20 µg/ml propidium iodide PI: 200 µl of 1000× (1 mg/ml) PI solution (appendix S12)
- distilled H_2O to 10 ml.

F11 KCl-HEPES hypotonic solution

- 30 mM KCl: 0.3 ml of 1 M KCl (appendix SS8)
- 5 mM HEPES: 50 µl of 1 M HEPES (free acid) (appendix SS2)
- pH adjusted to 7.5
- distilled H_2O to 10 ml.

F12 KCl-HEPES-T×100-PI solution

- 30 mM KCl: 0.3 ml of 1 M KCl (appendix SS8)
- 5 mM HEPES: 50 µl of 1 M HEPES (free acid) (appendix SS2)
- 0.9% Triton ×100: 90 µl of Triton ×100
- 135 µM propidium iodide PI: 900 µl of 1000× (1 mg/ml) PI solution (appendix S12)
- distilled H_2O to 10 ml.

F13 KCl-HEPES-PI solution

- 30 mM KCl: 0.3 ml of 1 M KCl (appendix SS8)
- 5 mM HEPES: 50 µl of 1 M HEPES (free acid) (appendix SS2)
- 45 µM propidium iodide PI: 300 µl of solution 1000× (1 mg/ml) PI (appendix S12)
- distilled H_2O to 10 ml.

F14 KCl-HEPES-MgSO₄ solution

- 50 mM KCl: 0.5 ml of 1 M KCl (appendix SS8)
- 5 mM HEPES: 50 µl of 1 M HEPES (free acid) (appendix SS2)
- 10 mM $MgSO_4$: 1 ml of 0.1 M $MgSO_4$ (appendix SS10)
- pH adjusted to 8
- distilled H_2O to 10 ml.

H Solutions for in situ hybridisation

H1 ribonuclease A solution

100 mg of ribonuclease A (type IA, Sigma R4875) are dissolved in 10 ml of H_2O and warmed at 100 °C in a water bath for 10 minutes to inactivate desoxyribonucleases. Store in 1 ml (10 mg/ml) aliquots at −20 °C.

H2 hybridisation medium

The following solutions are prepared:
- 50% dextran sulfate (Sigma D6001): 50 g in 100 ml of sterile H_2O
- sterile 20× SSC (appendix B4)
- deionised formamide (appendix M7)
- 1 M $NaH_2PO_4.H_2O$: 13.8 g in 100 ml of H_2O (final pH 6.8–7)
- 100× Denhardt solution = Ficoll 400 (Pharmacia): 0.5 g
                            polyvinylpyrrolidone: 0.5 g
                            bovine serum albumin (fraction V, Sigma A4503): 0.1 g
                            sterile H_2O to 25 ml
                            Store in 1 ml aliquots at −20 °C.

For 100 ml of **hybridisation medium** mix:
- 50 ml of deionised formamide (final concentration 50%, appendix M7)
- 10 ml of 20× SSC (final concentration 2×, appendix M4)
- 1 ml of 100× Denhardt solution (final concentration 1×)
- 4 ml of 1 M NaH_2PO_4 (final concentration 40 mM)

Filter through a 0.45 µm membrane, then add 20 ml of 50% dextran sulphate (final concentration 10%), adjust the pH to 7 if necessary and fill up to 100 ml with H_2O. Store in 1 ml aliquots at −20 °C.

H3 PBT solution

4 ml of a sterile 30% solution of bovine serum albumin (Sigma A3299) and 1 ml of Tween 20 (Sigma P1379) are added to 1000 ml of phosphate buffer (appendix B10) and the solution is filtered through a 0.22 µm membrane.

H4 anti-biotin antibody solution

anti-biotin antibody stock solution: 1 mg of goat antibiotin antibody raised in goat (Vector Biosys SP3000) is resuspended in 1 ml of sterile H_2O, centrifuged at 12500 g for 5 minutes and stored in 50 µl aliquots at −20 °C in the dark. *working solution*: 4 µl of the stock solution are diluted in 996 µl of PBT (appendix H3) and then kept on ice.

H5 antibody anti-goat IgG conjugated to fluorescein solution

stock solution of fluorescein conjugated anti-goat IgG antibody: 1 mg of the lyophilisated conjugated antigoat IgG antibody raised in rabbit (Nordic 3607) is resuspended in 2 ml of sterile H_2O, centrifuged at 12500 g for 5 minutes and stored in 100 µl aliquots at −20 °C in the dark.

working solution: 60 μl of stock solution are diluted in 940 μl of PBT (appendix H3) and then kept on ice in the dark.

H6 hybridisation medium for chromosomes in suspension

For 100 ml of hybridisation medium mix:
- 40 ml of deionised formamide (final concentration 40%, appendix M7)
- 20 ml of 20× SSC (final concentration 4×, appendix B4)
- 2 ml of 100× Denhardt solution (final concentration 2×, appendix H2)

The mixture is filtered through a 0.45 μm membrane, then 20 ml of 50% dextran sulphate are added (appendix H2, final concentration 10%). Fill up to 100 ml with distilled H_2O and store in 1 ml aliquots at −20 °C.

H7 hybridisation medium for polytene chromosomes

- deionised formamide (appendix M7): 60 μl
- DNA probe: 20 μl (1 μg)
- 50% dextran sulphate (Sigma D6001) (appendix H2): 20 μl
- 20× SSC (appendix B4): 20 μl.

H8 extravidine-peroxydase solution

- 10% bovine serum albumin (Sigma B2518) in 10× PBS⁻ (appendix B1): 50 μl
- distilled H_2O to 450 μl
- extravidine-peroxydase (Sigma E 2886): 1 μl

The 10% bovine serum albumin stock solution is stored at 4 °C.

H9 DAB (3,3'-Diaminobenzidine tetrahydrochloride) solution

- 0.5 mg/ml DAB solution (Polysciences): 1.5 ml
- 1% H_2O_2: 30 μl.

Since DAB is toxic, we recommend the use of powdered DAB dispensed in pill boxes to reduce handling. The necessary quantity of PBS⁻ (appendix B1) is introduced into the pill box to a final concentration of 0.5 mg/ml. The solution is stored in 1.5 ml aliquots (Eppendorf tubes) at −20 °C. Avoid freezing and thawing more than once and protect from light.

A stock solution of 30% H_2O_2 is distributed into aliquots in Eppendorf tubes and stored at −20 °C. The DAB solution is prepared and diluted just before use.

Special comment: all material which has been in contact with DAB must be immersed in a bleach bath for several hours and must be handled with gloves.

CM Culture media

CM1 basal media Ham F10, Ham F12, TC199 or RPMI1640

- 1× Ham F10, Ham F12, TC199 or RPMI1640 medium
- 100 U.I./ml penicillin, 100 μg/ml streptomycin, 0.25 μg/ml fungizone

Filter through a 0.22 μm membrane under sterile conditions and store at 4 °C.

CM2 MEM-10% FCS or MEM-5% FCS

- MEM minimum essential medium with Earle's salts and L-glutamine, free of $NaHCO_3$(Gibco)
- 2.2 g/litre sodium bicarbonate $NaHCO_3$
- 100 U.I./ml penicillin, 100 μg/ml streptomycin, 0.25 μg/ml fungizone
- 10% or 5% foetal calf serum FCS (Boerhinger Mannheim)
- final pH 7.3–7.4

Filter through a 0.22 μm membrane and keep at 4 °C.

CM3 TCM 199-HEPES-BSA

- 199 medium with Hank's salts, L-glutamine, 25 mM HEPES (Gibco 22350)
- 3 g/litre BSA fraction V
- 0.27 mM sodium pyruvate: 30 mg/litre
- gentamycin: 50 mg/litre

This medium is sterilised by filtration through a 0.22 μm membrane and can be stored for 1 month at 4 °C.

CM4 TCM 199-FCS-LH-FSH

- 199 medium with Earle's salts, L-glutamine, sodium bicarbonate (Sigma M4530)
- 20% foetal calf serum FCS
- 5 μg/ml luteinizing hormone LH (Sigma L7134 or LH: $1.0 \times$ NIH-LH-S1, Beckers et al., 1977)
- 0.5 μg/ml follicle stimulating hormone FSH (Biotech BURHS; or FSH: $75 \times$ NIH-FSH-S1, Beckers et al., 1977)
- 0.27 mM sodium pyruvate: 30 mg/ml
- 100 U.I./ml penicillin
- 100 μg/ml streptomycin

This medium is prepared in 5 ml aliquots the day before use, sterilised by filtration through a 0.22 μm membrane and used after incubation for one night in a CO_2 incubator.

CM5 TYB (Test Yolk Buffer) medium

- 1.0814 g of TES (N-tris(hydroxymethyl)methyl-2-aminoethane sulfonic acid)
- 256.7 mg of Tris [tris(hydroxymethyl)aminomethane]
- 0.5 g of dextrose
- 6.2 mg of streptomycin
- 3.7 mg of penicillin G
- 6.25 ml of fresh egg yolk
- 25 ml of distilled H_2O

The medium is centrifuged at $800 g$ for 10 minutes, the supernatant is discarded and the pH is adjusted to 7.4–7.5 with 0.1 M Tris (appendix SS14).

CM6 BWW (Biggers-Whitten-Whittingham) medium

stock medium: – 5.54 g of NaCl
– 356 mg of KCl
– 189 mg of $CaCl_2$
– 162 mg of KH_2PO_4
– 294 mg of $MgSO_4.7H_2O$
– 1 g of glucose
– 28 mg of sodium pyruvate
– 4.96 g of HEPES
– 25 mg of phenol red
– 5000 U.I. of penicillin
– 5 mg of streptomycin
– distilled H_2O to 1 litre

Filter through a 0.22 µm membrane. This medium can be stored for 15 days at 4 °C.

working medium (to be prepared extemporaneously):
– 100 ml of BWW stock medium
– 470 mg of lactic acid
– 210 mg of $NaHCO_3$
– 300 mg of human serum albumin

The pH is adjusted to 7.4 by bubbling with CO_2. Filter through a 0.22 µm membrane.

SS Stock solutions

SS1 0.1 M H_3BO_3 orthoboric acid

– 6.2 g of H_3BO_3
– H_2O to 1 litre

SS2 1 M HEPES

– 238.3 mg of HEPES (free acid)
– H_2O to 1 ml

SS3 10 mM ETB ethidium bromide

– 3.94 mg of ETB
– 1 ml of H_2O

Protect from light.

SS4 0.1 M EDTA

– 29.2 g of EDTA (ethylene diamine tetra acetic acid, anhydrous free acid)
– H_2O to 1 litre

SS5 0.5 M Na₂EDTA.2H₂O

Dissolve 186.1 g of $Na_2EDTA.2H_2O$ in 700 ml of H_2O, adjust the pH to 8 with about 50 ml of 10 M NaOH (appendix SS12) and add H_2O to 1 litre.

SS6 0.1 M EGTA

- 3.8 g of EGTA (ethylene glycol-bis(beta-aminoethylether)-tetra acetic acid, free acid)
- H_2O to 100 ml

SS7 1 M CaCl₂

- 147 g of $CaCl_2.2H_2O$
- H_2O to 1 litre

SS8 1 M KCl

- 74.6 g of KCl
- H_2O to 1 litre

SS9 1 M MgCl₂

- 20.3 g of $MgCl_2.6H_2O$
- H_2O to 100 ml

SS10 1 M MgSO₄

- 24.6 g of $MgSO_4.7H_2O$
- H_2O to 100 ml

SS11 1 M NaCl

- 58.4 g of NaCl
- H_2O to 1 litre

SS12 1 M and 5 M NaOH

1 M NaOH
- 40 g of NaOH
- H_2O to 1 litre
5 M NaOH
- 200 g of NaOH
- H_2O to 1 litre

SS13 1 M Na₂HPO₄

- 142 g of anhydrous Na_2HPO_4
- H_2O to 1 litre

SS14 1 M and 0.1 M Tris

Dissolve 121 g (1 mole) or 12.1 g (0.1 mole) of Tris base in 800 ml of H_2O, adjust to the appropriate pH with concentrated HCl or CH_3COOH and add H_2O to 1 litre.

SS15 0.1 M Na_3citrate.$2H_2O$

- 29.4 g of Na_3citrate.$2H_2O$
- H_2O to 1 litre

B Buffer solutions

B1 1× PBS⁻ Dulbecco's Ca^{++} and Mg^{++} free phosphate buffered saline

- 137 mM NaCl: 8 g
- 2.7 mM KCl: 0.2 g
- 8 mM Na_2HPO_4.$12H_2O$: 2.89 g
- 1.47 mM KH_2PO_4: 0.2 g
- H_2O to 1 litre
- final pH 7.2

Autoclave. A solution 10× PBS⁻ can be prepared, the pH is adjusted to 7.2–7.4 with 1 M NaOH (appendix SS12).

B2 BSS⁺ Hank's balanced salt solution with Ca^{++}

- 137 mM NaCl: 8 g
- 5.4 mM KCl: 0.4 g
- 0.81 mM $MgSO_4$.$7H_2O$: 0.2 g
- 0.33 mM Na_2HPO_4.$12H_2O$: 0.12 g
- 0.44 mM KH_2PO_4: 0.06 g
- H_2O to 1 litre.

After autoclaving add:
- 6 ml of a 5.6% $NaHCO_3$ solution sterilised by filtration
- 2 ml of sterile 1 M $CaCl_2$ solution (appendix SS7)
- final pH 7.2–7.4.

B3 Sorensen's buffer pH 6.8

Prepare two solutions **X** and **Y**:
X = 1/15 M KH_2PO_4: 9.08 g in 1 litre of H_2O
Y = 1/15 M Na_2HPO_4.$12H_2O$: 23.88 g in 1 litre of H_2O
For 1 litre of Sorensen's solution final pH 6.8, mix 508 ml **X** + 492 ml **Y**.

B4 20× SSC standard saline solution

- 3 M NaCl: 175.2 g
- 0.3 M Na_3citrate.$2H_2O$: 88.2 g
- H_2O to 1 litre and autoclave

For 2× or 1× SSC (pH 7) washing solutions used in the protocol of *in situ* hybridisation or for staining, prepare a 20× SSC solution the pH of which is adjusted to 6.3 with HCl. For the 50% formamide 2× SSC (pH 7) solutions used in the protocol of *in situ* hybridisation, prepare a 20× SSC solution the pH of which is adjusted to 5.8 with HCl.

B5 10× TE

- 100 mM Tris: 1.21 g
- 10 mM EDTA: 0.292 g
- final pH adjusted to 8 with 1 N HCl
- H_2O to 100 ml.

B6 MacIlvaine's buffer pH 7

Prepare two solutions **X** and **Y**:
X = 0.1 M citric acid $C_6H_8O_7.H_2O$: 21 g in 1 litre of H_2O
Y = 0.2 M $Na_2HPO_4.2H_2O$: 35.6 g in 1 litre of H_2O
For 1 litre of MacIlvaine's solution at a final pH of 7, mix 181.5 ml **X** + 818.5 ml **Y**.

B7 borate buffer pH 9.2

- 0.1 M Na_2SO_4: 14.2 g
- 0.0095 M $Na_2B_4O_7.10H_2O$: 3.62 g
- H_2O to 1 litre
- final pH 9.2.

B8 borate buffer pH 8.4

Prepare two solutions **A** and **B**:
A = 0.2 M boric acid H_3BO_3: 12.36 g in 1 litre of H_2O
B = 0.05 M borax $Na_2B_4O_7.10H_2O$: 19.07 g in 1 litre of H_2O
For 1 litre of solution at a final pH of 8.4, mix 250 ml **A** + 57.5 ml **B** + H_2O to 1 litre

B9 10× Earle's bicarbonate free solution

- 1.16 M NaCl: 68 g
- 0.054 M KCl: 4 g
- 0.01 M $NaH_2PO_4.H_2O$: 1.4 g
- 8.1 mM $MgSO_4.7H_2O$: 2 g
- 0.055 M glucose: 10 g
- 18 mM $CaCl_2$: 2 g
- phenol red
- H_2O to 1 litre

After dilution, the pH of 1× Earle's solution is adjusted to 5.3 or 6.5 with saturated Na_2HPO_4 solution.

B10 phosphate buffer

Prepare two solutions X and Y.
$X = 0.2$ M $NaH_2PO_4.2H_2O$: 6.24 g in 200 ml of H_2O
$Y = 0.2$ M $Na_2HPO_4.12H_2O$: 71.63 g in 1 litre of H_2O
For 2 litres of phosphate buffer at a final pH of 7.2–7.3, mix 190 ml X + 810 ml Y + 1 litre of H_2O.
Add 17.4 g of NaCl (final concentration 0.149 M) and filter through a 0.22 μm membrane.

B11 1× TBS

– 10 mM Tris (Tris base): 1.21 g
– 100 mM NaCl: 5.84 g
– final pH adjusted to 7.5–7.6 with 1 N HCl
– distilled H_2O to 1 litre.

B12 Ringer solution

– 128 mM NaCl: 7.5 g
– 4.7 mM KCl: 0.35 g
– 1.42 mM $CaCl_2.2H_2O$: 0.21 g
– distilled H_2O to 1 litre
Adjust the pH to 7.4–7.5 with $NaHCO_3$ solution, autoclave and store at 4 °C.

V Solutions for in vivo treatments

V1 10% DMSO-foetal calf serum

– foetal calf serum: 90 ml
– sterile DMSO (Sigma D2650): 10 ml
This solution is stored at 4 °C.

V2 1× colchicine solution

10× stock solution (40 μg/ml): 0.4 mg of colchicine (Sigma C9754) in 10 ml of PBS⁻ (appendix B1).
Store in 1 ml aliquots at −20 °C.
1× working solution (4 μg/ml): dilute 1 ml of the 10× solution in 9 ml of PBS⁻ and store in 1 ml aliquots at −20 °C.

V3 1× thymidine solution

1× working solution (10 mg/ml = 41.3 mM): dissolve 100 mg of thymidine (Sigma T1895) in 10 ml of MEM-5% FCS medium (appendix CM2). Filter under sterile conditions through a 0.22 μm membrane. The solution can be stored for a week at 4 °C.

V4 1× amethopterin solution

100× stock solution (10^{-3} M): dissolve 1 mg of amethopterin (Sigma M8407) in 2.2 ml of PBS⁻ (appendix B1).
Store in 100 μl aliquots at −20 °C.
1× working solution (10^{-5} M): dilute 100 μl of the 100× solution in 9 ml of PBS, filtered through a 0.22 μm membrane and store in 1 ml aliquots at −20 °C.

V5 1× BrdU solution (5-bromo-2′-deoxyuridine)

1× working solution (3.07 mg/ml): 30.7 mg of BrdU (Sigma B5002) in 10 ml of PBS⁻ (appendix B1). Filter through a 0.22 μm membrane and store in 250 or 500 μl aliquots at −20 °C.

V6 1× FdU solution (5-fluoro-2′-deoxyuridine)

10× stock solution (3 mM): 7.5 mg of FdU (Sigma F0503) in 10 ml of PBS⁻ (appendix B1). Store in 1 ml aliquots at −20 °C.
1× working solution (75 μg/ml = 0.3 mM): dilute 1 ml of the 10× solution in 9 ml of PBS⁻, filter through a 0.22 μm membrane and store in 0.5 or 1 ml aliquots at −20 °C.

V7 10× 5-azacytidine (5-ACR) solution

10× stock solution (10^{-4} M): dissolve 24.42 mg of 5-ACR in 10 ml of distilled H₂O.
Filter through a 0.22 μm membrane and store in 1 ml aliquots at −20 °C.

V8 10× 5-deoxyazacytidine (5-dACR) solution

10× stock solution (10^{-4} M): dissolve 22.82 mg of 5-dACR in 10 ml of distilled H₂O.
Filter through a 0.22 μm membrane and store in 1 ml aliquots at −20 °C.

Glossary

Achromatic spindle   A structure occurring during mitosis or meiosis and playing a role in distribution of the chromatids or of the chromosomes (diacinese) in daughter cells.

Acrocentric chromosome   A chromosome with a nearly terminal centromere.

Alternative and adjacent orientation   In reciprocal translocation, a quadrivalent is formed during meiosis, between the chromosomes involved in this abnormality and their homologous. An alternative orientation in the quadrivalent allows the migration of a balanced chromosome complement to opposite poles. In contrast, in adjacent orientation each pole receives an unbalanced chromosome complement.

Aneuploidy   A condition of a cell in which the chromosome number is not an exact multiple of the normal haploid set for the species. This situation is due to a loss or, in contrast, an addition of one or several chromosomes.

Antibody   A protein produced by an organism in response to an antigen presence capable of coupling specifically with it.

Antigen   A foreign substance that, upon introduction into an organism, stimulates the production of antibodies.

Basic number   The number of chromosomes characterising a species.

Biotinylation   Incorporation of a nucleotide biotin-coupled in a DNA probe.

Bivalent   A structure formed by two homologous chromosomes pairing during meiosis.

Cell chimerism   The presence in an organism, of a mixture of two or several genetically different cells coming from different zygotes.

Cellular mosaic   An organism composed of two or more cell lines, these cell lines being derived from the same zygotic origin. These cell lines are of different characters which are transmissible by cell heredity, e.g. the chromosome number.

Centromere   The region of each chromosome with which the achromatic spindle is associated with the chromosome.

Centromeric index   Relation between the short arm and the total length of the chromosome.

Chromosome pairing   The association of homologous chromosomes during meiosis, and exceptionally during mitosis (drosophylla).

Colchicine   An alkaloid extracted from colchicum, that inhibits the formation of the achromatic spindle within a cell division by the tubulins combination.

Complementary DNA (cDNA)   Simple-sequence DNA produced from a RNA template by the enzymatic action in vitro.

Constitutive heterochromatin   Chromosome structures different from the euchromatin by their biochemical composition and by several cytological characteristics. It essentially consists in repetitive DNA sequences, often hypermethylated.

Deletion   The loss of a chromosome segment without centromere after break.

DNA denaturation   Reversible dissociation of the two strands of the DNA molecule.

Double stranded DNA   DNA molecule consisting of two complementary strands linked by low chemical linkages (non covalent).

Duplication   The doubling of a chromosome or of a chromosome fragment which is isolated or is attached to another chromosome. This leads to an allelic desequilibrium and to a gene dosage desequilibrium (trisomy).

Euploidy   Condition of a cell containing whole number of haploid chromosome sets.

Facultative heterochromatin   Chromosome segments or chromosomes consisting in euchromatin becoming condensed during certain stages of the cell cycle. They may be different, as e.g. the inactive chromosome X of the mammals female diploid cells.

Fundamental number   The total number of chromosome arms.

Haploidy   In eukaryotes, the haploidy correspond to one set of chromosomes. It is the natural condition after the meiotic reduction.

Homoeologous chromosomes   Chromosomes that present the same or almost the same banding patterns, but which are present in different species.

Homologous chromosomes   Chromosomes that pair during meiosis. Each homologue is a duplicate of mother or father chromosome. Homologous chromosomes contain the same genes.

Hypodiploidy   Absence of one or several chromosomes in a diploid chromosome set.

Inversion   Inverted chromosome segment further to a double break; the inversion is paracentric if the centromere is not located on the inverted segment and pericentric if it contains a centromere.

Isochromosome   A chromosome with genetically identical arms, attached to a central centromere in symmetric position.

Karyograme   Chromosome graphic representation of a specie obtained by measuring several cells and considered as representative for this specie.

Karyotype   A chromosome complement of one cell, classified in a decreasing order, according to their size and the centromere position.

Kilobase   A unit of length of a nucleic acid equal to a sequence of 1000 bases.

Lampbrush chromosomes   A special type of chromosome in primary oocyte nuclei, both vertebrate and invertebrate, during the prophase of the first meiotic division. At this stage, the DNA of the four chromatids is decomposed projecting in loops that extend laterally along the chromosome axis, that is the lampbrush aspect. Each loop contains an important transcriptional activity. These chromosomes have been observed in amphibians and in birds until now.

Metacentric chromosomes   A chromosome with a median centromere.

Mitotic index   Proportion of cells in division.

Monosomy   Presence of a chromosome in one exemplary in a diploid chromosome set.

Non disjunction   Accidental migration of two homologues to the same pole, instead of their migration to opposite poles. The two resulting sister cells are respectively trisomic and monosomic for this chromosome, for a diploid mother cell.

Nucleic bases complementarity   Specific pairing (AT and GC) of nucleic bases.

Polyploidy   Presence of three, four or more haploid chromosome lots in a cell, tissue or organism.

Polythene chromosome   A giant chromosome consisting of many chromatids lying in parallel and producing further several DNA replication cycles without cell division. The DNA is condensed in certain regions along the chromatin (dark bands) and because of these many chromatids, a bands pattern is visible along the chromosome axis. The polythene chromosomes are found only in some species, e.g. in Drosophila salivary gland.

Pseudodiploidy   Presence of a diploid chromosome set in a normal chromosome complement cell, tissue or organism.

Quadrivalent   A meiotic association of four chromosomes, e.g. because of a reciprocal translocation.

Reciprocal translocation   Chromosome rearrangement that results in two breaks with an interchange of chromosomal segments between nonhomologous chromosomes and produces two monocentric rearranged chromosomes.

Recombination index   The sum of the haploid chromosome number of a specie and the average number of chiasmata per nucleus.

Relative length   Chromosome length (in percent) of the total length of all cell chromosomes.

ribosomal RNA   RNA molecules which make up a part of a ribosome's content.

Robertsonian series   Related species having the same fundamental number but different chromosome number because of robertsonian translocations.

Robertsonian translocation   Centromeric regions fusion of two acrocentric chromosomes.

Satellite DNA   The DNA fraction different in base composition (density and centrifugation) from the major DNA component. It consists in repetitive sequences.

Simple-stranded DNA   DNA molecule consisting in one strand DNA.

Submetacentric chromosome   A chromosome whose centromere is located in sub-median position.

Synaptonemal complex   A tripartite protein complex appearing in meiotic prophase, at pachytene stage, in which occur the meiotic crossings-over (meiotic re combinations).

Tandem translocation   The fusion "end to end" of two chromosomes, with loss of one centromere.

Tetraploidy   Cell or organism having four chromosomal haploid lots.

Tetravalent   see quadrivalent.

Translocation   Fragments exchange between two chromosomes.

Triploidy   Organism having three haploid sets of chromosomes, which may result from a double in vitro fertilisation of an oocyte, or from a normal fertilisation of a diploid oocyte.

Trisomy   Presence of three chromosomes in a diploid chromosome set.

Trivalent   Three meiotic-associated chromosomes in a triploid or trisomic organism. It is observed also in meiosis, in heterozygosity for a robertsonian translocation.

True hermaphrodite   An organism or a species carrying male and female tissues. In mammals they are sterile.

Zygote   In Eucaryotes, a diploid cell formed by the fusion of two gametes.

References

Almeida Toledo L.F., Viegas-Péquignot E., Foresti F., Toledo Filho S.A. et Dutrillaux B. (1988). BrdU replication patterns demonstrating chromosome homoeologies in two fish species, genus *Eigenmannia*. *Cytogenet. Cell Genet.*, 48, 117–120.

Almeida P.A. et Bolton V.N. (1993). Immaturity and chromosomal abnormalities in oocytes that fail to develop pronuclei following insemination in vitro. *Hum. Reprod.*, 8, 229–232.

Ambros P.F. et Sumner A.T. (1987). Correlation of pachytene chromomeres and metaphase bands of human chromosomes, and distinctive properties of telomeric regions. *Cytogenet. Cell Genet.*, 44, 223–228.

Angelier N. et Lacroix J.-C. (1975). Complexes de transcription d'origines nucléolaire et chromosomique d'ovocytes de *Pleurodeles waltlii* et *P. poireti* (Amphibiens, Urodèles). *Chromosoma*, 51, 323–335.

Ansari H.A., Hediger R., Fries R. et Stranzinger G. (1988). Chromosomal localization of the major histocompatibility complex of the horse (ELA) by in situ hybridization. *Immunogenetics*, 28, 362–364.

Ansari H.A., Bosma A.A., Long S.E. (coordinator) et Popescu C.P. (1994). Clarification of chromosome nomenclature in the sheep (*Ovis aries*): Report of the committee for the standardization of the sheep karyotype. *Cytogenet. Cell Genet.*, 67, 114–115.

Apiou F., Rumpler Y., Warter S., Vezuli A. et Dutrillaux B. (1996). Demonstration of homoeologies between human and lemur chromosomes by chromosome painting. *Cytogenet. Cell Genet.*, 72, 50–52.

Ashburner M. (1989). Drosophila, a laboratory manual. Cold Spring Harbor Laboratory Press.

Balbiani E.G. (1881). Sur la structure du noyau des cellules salivaires chez les larves de *Chironomus*. *Zool. Anz.*, 4, 637–641.

Barbin A., Montpellier C., Kokalj-Vokac N., Gibaud A., Niveleau A., Malfoy B., Dutrillaux B. et Bourgeois C.A. (1994). New sites of methylcytosine-rich DNA detected on metaphase chromosomes. *Hum. Genet.*, 94, 684–692.

Bartholdi M.F., Ray F.A., Jett J.H., Cram L.S. et Kraemer P.M. (1984). Flow karyotyping of serially cultured Chinese hamster cell lineages. *Cytometry*, 5, 534–538.

Bartholdi M.F., Meyne J., Johniston R.G. et Cram L.S. (1989). Chromosome band analysis by slit-scan flow cytometry. *Cytometry*, 10, 124–133.

Barths J. (1987). Durchflusszytometrische und cytogenetische Charakterisierung von permanenten Zellinien des Chinesischen Hamsters (Cricetulus griseus). PhD Thesis, Faculty of Biology, University of Kaiserslautern, Allemagne.

Beckers J.-F., Closset J., Maghuin-Rogister G. et Hennen G. (1977). Bovine follitropin: isolation and characterization of the native hormone and its a and b subunits. *Biochimie*, 59, 825–831.

Benkhalifa M., Bonneau M., Popescu P., Boscher J., Boucher D. et Malet P. (1992). A method for cytogenetic analysis of boar spermatozoa using hamster oocytes. *Ann. Genet.*, 35, 61–64.

Bernardi G., Olofsson B., Filipski J., Zerial M., Salinas J., Cuny G., Meunier-Rotival M. et Rodier F. (1985). The mosaic genome of warm-blooded vertebrates. *Science*, 228, 953–958.

Bickmore W.A. et Sumner A.T. (1989). Mammalian chromosome banding – an expression of genome organization. *Trends Genet.*, 5, 144–148.

Biémont C. et Gautier C. (1988). Localization and polymorphism of mdg-1, copia, I and P mobile elements in genomes of *Drosophila melanogaster*, from data of inbred lines. *Heredity*, 60, 335–346.

Blackley R.L. (1969). In: The biochemistry of folic acid and related pteridines, Frontiers of Biology, 13, Neuberger A. et Tatum E.L. (eds), North Holland Publishing Company, Amsterdam.

Blaise F., Aycardi J., Boscher J. et Popescu C.P. (1990). Flow cytometry of normal and aberrant pig karyotypes. *Ann. Genet.*, 33, 146–151.

Blumenthal A.B., Dieden J.D., Kapp L.N. et Sedat J.W. (1979). Rapid isolation of metaphase chromosomes containing high molecular weight DNA. *J. Cell Biol.*, 81, 255–259.

Bobrow M., Madan K. et Pearson P.L. (1972). Staining of some specific regions of human chromosomes, particularly the secondary constriction of n°9. *Nature*, 238, 122–128.

Bonnanfant-Jaïs M.L. et Mentré P. (1983). Study of oogenesis in the newt *Pleurodeles waltlii* M. *J. Submicrosc. Cytol.*, 15, 453–478.

Bonner W.A., Hulett H.R., Sweet R.G. et Herzenberg L.A. (1972). Fluorescence activated cell sorting. *Rev. Sci. Instr.*, 43, 404–409.

Boschman G.A., Rens W., Manders E.M.M., van Oven C.H., Barendsen G.W. et Aten J.A. (1990). On-line sorting of human chromosomes by centromeric index, and identification of sorted populations by GTG-banding and fluorescent in situ hybridization. *Hum. Genet.*, 85, 41–48.

Boschman G.A., Rens W., van Oven C.H., Manders E.M.M. et Aten J.A. (1991). Bivariate flow karyotyping of human chromosomes: Evaluation of variation in Hoechst 33258 fluorescence, Chromomycin A3 fluorescence and relative chromosomal DNA content. *Cytometry*, 12, 559–569.

Bostock C.J. et Sumner A.T. (1978). Substructure of chromosomes and the banding phenomena. In: The eukaryotic chromosome, North Holland Publishing Company, Amsterdam, New York and Oxford, pp 375–405.

Boyum A. (1968). Separation of leucocytes from blood and bone marrow. *Scand. J. Clin. Lab. Invest.*, 21, (suppl. 97), 31.

Breneman J.W., Swiger R.R., Ramsey M.J., Minkler J.L., Eveleth J.G., Langlois R.A. et Tucker J.D. (1995). The development of painting probes for dual-color and multiple chromosome analysis in the mouse. *Cytogenet. Cell Genet.*, 68, 197–202.

Bridges C.B. (1935). Salivary chromosome maps with a key to the banding of the chromosomes of *Drosophila melanogaster*. *J. Hered.*, 26, 60–64.

Bridges C.B. (1937). Correspondences between linkage maps and salivary chromosome structure as illustrated in the tip of chromosome 2R of *Drosophila melanogaster*. *Cytologia (Fujii Jubil. Vol.)*, 745–755.

Bridges C.B. (1938). A revised map of the salivary gland X-chromosome of *Drosophila melanogaster*. *J. Hered.*, 29, 11–13.

Bridges C.B. et Bridges P.N. (1939). A new map of the second chromosome. A revised map of the right limb of second chromosome of *Drosophila melanogaster*. *J. Hered.*, 30, 475–476.

Bridges P.N. (1941a). A revised map of the left limb of the third chromosome of *Drosophila melanogaster*. *J. Hered.*, 32, 64–65.

Bridges P.N. (1941b). A revision of the salivary gland 3R-chromosome map. *J. Hered.*, 32, 299–300.

Bridges P.N. (1942). A new map of the salivary gland 2L-chromosome of *Drosophila melanogaster*. *J. Hered.*, 33, 403–408.

Broccoli D. et Cooke H. (1993). Aging, healing and the metabolism of telomeres. *Am. J. Hum. Genet.*, 52, 657–660.

Burkholder G.D. (1993). The basis of chromosome banding. *Appl. Cytogenet.*, 19, 181–186.

Burkin D.J., Broad T.E. et Jones C. (1996). The chromosomal distribution and organization of sheep satellite I and II centromeric DNA using characterized sheep-hamster somatic cell hybrids. *Chromos. Res.*, 4, 49–55.

Callan H.G. (1963). The nature of lampbrush chromosomes. *Int. Rev. Cytol.*, 15, 1–34.

Callan H.G. (1986). Lampbrush chromosomes. *Mol. Biol. Biochem. Biophys.*, 36, 1–254.

Callan H.G. et Lloyd L. (1956). Visual demonstration of allelic differences within cell nuclei. *Nature*, 178, 355–357.

Callan H.G. et Lloyd L. (1960). Lampbrush chromosomes of crested newts *Triturus cristatus* (Laurenti). *Phil. Trans. Roy. Soc. London Biol.*, 243, 135–219.

Callan H.G. et Lloyd L. (1975). Working maps of the lampbrush chromosomes of Amphibian. In: Handbook of genetics, King R.C. (ed), New York, Plenum, 4, pp 57–77.

Carrano A.V., Gray J.W., Moore D.H., Minler J.L., Mayall B.H., Van Dilla M.A. et Mendelsohn M.L. (1976). Purification of the chromosomes of the Indian muntjac by flow sorting. *J. Histochem. Cytochem.*, 24, 348–354.

Carrano A.V., Gray J.W., Langlois R.G., Burkhardt-Schultz K.J. et Van Dilla M.A. (1979). Measurement and purification of human chromosomes by flow cytometry and sorting. *Proc. Natl. Acad. Sci.*, 76, 1382–1384.

Caspersson T., Farber S., Foley G.E., Kudynowski J., Modest E.J., Simonsson E., Wagh V. et Zech L. (1968). Chemical differentiation along metaphase chromosomes. *Exp. Cell Res.*, 49, 219–222.

Celeda D., Bettag U. et Cremer C. (1992). PCR amplification and simultaneous digoxigenin incorporation of long DNA probes for fluorescent in situ hybridization. *Biotechniques*, 12, 98–102.

Celeda D. (1993). Kombination von DNA Proben Präparation mittels Polymerase Chain Reaction und in situ Hybridisierung an menschlichen Chromosomen unter Verwendung von PCR Puffer Systemen und physiologischer Kochsalzlösung. PhD Thesis, Faculty of Biology, University of Heidelberg, Allemagne.

Celeda D., Aldinger K., Haar F.M., Hausmann M., Durm M., Ludwig H. et Cremer C. (1994). Rapid fluorescence in situ hybridization with repetitive DNA probes: Quantification by digital image analysis. *Cytometry*, 17, 13–25.

Chérif D., Bernard O. et Berger R. (1989). Detection of single-copy genes by nonisotopic in situ hybridization on human chromosomes. *Hum. Genet.*, 81, 358–362.

Chérif D., Julier C., Delattre O., Derré J., Lathrop G.M. et Berger R. (1990). Simultaneous localization of cosmids and chromosome R-banding by fluorescence microscopy: application to regional mapping of human chromosome 11. *Proc. Natl. Acad. Sci.*, 87, 6639–6643.

Chowdhary B.P., Harbitz I., Davies W. et Gustavsson I. (1991). Chromosomal localization of the glucose phosphate isomerase (GPI) gene in cattle, sheep and goat by in situ hybridization: chromosomal banding homology versus molecular conservation in Bovidae. *Hereditas*, 114, 161–170.

Comings D.E. (1973). Biochemical mechanisms of chromosome banding and color banding with acridine orange. In: Chromosome identification – Techniques and applications in Biology and Medecine, Caspersson T. and Zeck L. (eds), New York, Academic Press, pp 292–306.

Comings D.E. (1978). Mechanisms of chromosome banding and implications for chromosome structure. *Ann. Rev. Genet.*, 12, 25–46.

Comings D.E., Avelino E., Okada T.A. et Wyandt H.E. (1973). The mechanism of C- and G-banding of chromosomes. *Exp. Cell Res.*, 77, 469–493.

Committee on the standardized genetic nomenclature for mice (1972). Standard karyotype of the mouse, *Mus Musculus*. *J. Hered.*, 63, 69–72.

Committee for a standardized karyotype of *Rattus norvegicus* (1973). Standard karyotype of the Norway rat, *Rattus norvegicus. Cytogenet. Cell Genet.*, 12, 199–205.

Committee for the standardized karyotype of the domestic pig (1988). Standard karyotype of the domestic pig. *Hereditas*, 109, 151–157.

Cooke H.J. et Hindley J. (1979). Cloning of human satellite DNA, different components are in different chromosomes. *Nucl. Ac. Res.*, 10, 3177–3197.

Counce S. et Meyer G. (1973). Differentiation of the synaptonemal complex and the kinetochore in Locusta spermatocytes studied by whole mount microscopy. *Chromosoma*, 44, 231–253.

Cremer C., Gray J.W. et Ropers H.H. (1982). Flow cytometric characterization of a Chinese hamster X man hybrid cell line retaining the human Y chromosome. *Hum. Genet.*, 60, 262–266.

Cremer C., Rappold G., Gray J.W., Müller C.R. et Ropers H.H. (1984). Preparative dual beam sorting of the human Y chromosome and in situ hybridization of cloned DNA probes. *Cytometry*, 5, 572–579.

Cremer C., Hausmann M., Zuse P., Aten J.A., Barths J. et Bühring H.-J. (1989a). Flow cytometry of chromosomes: Principles and applications in medecine and molecular biology. *Optik*, 82, 9–18.

Cremer C., Dölle J., Hausmann M., Bier F.F. et Rohwer P. (1989b). Laser in cytometry: Applications in flow cytogenetics. *Ber. Bunsenges. Phys. Chem.*, 93, 327–335.

Cremer C. et Cremer T. (1992). Analysis of chromosomes in molecular tumor and radiation cytogenetics: approaches, applications, perspectives. *Eur. J. Histochem.*, 36, 15–25.

Cremer T., Landegent J.E., Brückner H., Scholl H.P., Schardin M., Hager H.D., Devilee P., Pearson P.L. et van der Ploeg M. (1986). Detection of chromosome aberrations in the human interphase nucleus by visualization of specific target DNAs with radioactive and non-radioactive in situ hybridization techniques: diagnosis of trisomy 18 with probe L1.84. *Hum. Genet.*, 745, 346–352.

Cremer T., Lichter P., Borden J., Ward D.C. et Manuelidis L. (1988). Detection of chromosome aberrations in metaphase and interphase tumor cells by in situ hybridization using chromosome specific library probes. *Hum. Genet.*, 80, 235–246.

Darzynkiewicz Z. et Crissman H.A. (eds) (1990). Flow cytometry. *Meth. Cell Biol.*, 33, Academic Press, San Diego.

Denver Conference (1960). A proposed standard system of nomenclature of human mitotic chromosomes. *Lancet*, i, 1063–1065.

van Dilla M.A. et Deaven L.L. (1990). Construction of gene libraries for each human chromosome. *Cytometry*, 11, 208–218.

Disteche C.M., Carrano A.V., Ashworth L.K., Burkhardt-Schultz K. et Latt S.A. (1981). Flow sorting of the mouse Cattanach X-chromosome, t(X;7) 1Ct, in an active and inactive state. *Cytogenet. Cell Genet.*, 29, 189–197.

Dittrich W. et Göhde W. (1969). Impulsfluorometrie bei Einzelzellen in Suspension. *Z. Naturforsch.*, 24b, 360–361.

Dixon S.C., Miller N.G.A., Carter N.P. et Tucker E.M. (1992). Bivariate flow cytometry of farm animal chromosomes: a potential tool for gene mapping. *Anim. Genet.*, 23, 203–210.

Dölle J., Hausmann M. et Cremer C. (1991). Background and peak evaluation of one parameter flow karyotypes on a PC/AT computer. *Analyt. Cell Pathol.*, 3, 119–132.

Dölle J. (1994). Echtzeitanalyse von Metaphase-Chromosomen für die Sortierung im Slit-scan-flussphotometer. PhD. Thesis, Faculty of Natural Sciences, University of Heidelberg.

Drouin R., Messier P-E. et Richer C-L. (1989). DNA denaturation for ultrastructural banding and the mechanism underlying the fluorochrome-photolysis-Giemsa technique studied with anti-5-bromodeoxyuridine antibodies. *Chromosoma*, 98, 174–180.

Dudin G., Cremer T., Schardin M., Hausmann M., Bier F. et Cremer C. (1987). A method for nucleic acid hybridization to isolated chromosomes in suspension. *Hum. Genet.*, 76, 290–292.

Dudin G., Steegmayer E.W., Vogt P., Schnitzer H., Diaz E., Howell K.E., Cremer T. et Cremer C. (1988). Sorting of chromosomes by magnetic separation. *Hum. Genet.*, 80, 111–116.

Dutrillaux B. (1973). Nouveau système de marquage chromosomique: les bandes T. *Chromosoma*, 41, 395–402.

Dutrillaux B. (1975a). Traitements discontinus par le BrdU et coloration par l'acridine orange: obtention de marquages R, Q et intermedias. *Chromosoma*, 52, 261–273.

Dutrillaux B. (1975b). Sur la nature et l'origine des chromosomes humains. Monog. Ann. Génétique. Expansion Scientifique Française, Paris.

Dutrillaux B. (1979). Chromosomal evolution in primates: tentative phylogeny from *Microcebus murimus* (Prosimian) to man. *Hum. Genet.*, 48, 251–314.

Dutrillaux B. et Lejeune J. (1971). Sur une nouvelle technique d'analyse du caryotype humain. *C. R. Acad. Sci. (Paris)*, 273, 2638–2640.

Dutrillaux B. et Covic M. (1974). Etude de facteurs influençant la dénaturation thermique ménagée des chromosomes. *Exp. Cell Res.*, 85, 143–153.

Dutrillaux B. et Viegas-Péquignot E. (1981). High resolution of R- and G-banding on the same preparation. *Hum. Genet.*, 57, 93–95.

Dutrillaux B. et Couturier J. (1981). La pratique de l'analyse chromosomique. Masson, Paris, pp 7–86.

Dutrillaux B., Finaz C., de Grouchy J. et Lejeune J. (1972). Comparison of banding patterns of human chromosomes obtained with heating, fluorescence, and proteolytic digestion. *Cytogenetics*, 11, 113–116.

Dutrillaux B., Laurent C., Couturier J. et Lejeune J. (1973). Coloration par l'acridine orange de chromosomes préalablement traités par le 5-bromodéoxyuridine (BUdR). *C. R. Acad. Sci.*, 276, 3179–3181.

Dutrillaux B., Couturier J., Lombard M. et Chauvier G. (1979a). Cytogénétique de deux Lorisidae (*Nycticebus coucang* et *Perodicticus potto*). Comparaison avec les lémuriens et les simiens. *Ann. Génét.*, 22, 93–98.

Dutrillaux B., Aurias A. et Lombard M. (1979b). Présence de chromosomes communs chez un rongeur (*Eliomys quercinus*, Lérot) et chez les primates. *Ann. Génét.*, 22, 21–24.

Echard G., Yerle M., Gellin J., Dalens M. et Gillois M. (1986). Assignment of the major histocompatibility complex to the p1.4–q1.2 region of chromosome 7 in the pig (*Sus scrofa domestica* L.) by in situ hybridization. *Cytogenet. Cell Genet.*, 41, 126–128.

Ectors F.J., Koulischer L., Jamar M., Herens C., Verloes A., Remy B. et Beckers J.-F. (1995). Cytogenetic study of bovine oocytes matured in vitro. *Theriogenology*, 44, 445–450.

Engels W.R., Preston C.R., Thompson P. et Eggleston W.B. (1986). In situ hybridization to Drosophila salivary chromosomes with biotinylated DNA probes and alkaline phosphatase. *Focus*, 8, 6–8.

van den Engh G., Trask B., Cram S. et Bartholdi M. (1984). Preparation of chromosome suspension for flow cytometry. *Cytometry*, 5, 108–117.

van den Engh G., Hanson D. et Trask B. (1990). A computer program for analysing bivariate flow karyotypes. *Cytometry*, 11, 173–183.

Evans E.P., Breckon G. et Ford C.E. (1974). An air drying method for meïotic preparation from mammalian testis. *Cytogenetics*, 3, 289–294.

Fan Y., Davis L.M. et Shows T.B. (1990). Mapping small DNA sequences by fluorescence in situ hybridization directly on banded chromosomes. *Proc. Natl. Acad. Sci.*, 87, 6223–6227.

Feinberg A.P. et Vogelstein B. (1983). A technique for radiolabelling DNA restriction endonuclease fragments to high specific activity. *Anal. Bioch.*, 132, 6–13.

Ferguson-Smith M.A. et Page B.M. (1973). Pachytene analysis in human reciprocal (10;11) translocation. *J. Med. Genet.*, 10, 283–286.

Flannery A.V. et Hill R.S. (1988). The effect of heat-shock on the morphology of amphibian lampbrush chromosomes. *Exp. Cell Res.*, 177, 9–18.

Fries R., Hediger R. et Stranzinger G. (1986). Tentative chromosomal localization of the bovine major histocompatibility complex by in situ hybridization. *Anim. Genet.*, 17, 287–294.

Funaki K., Matsui S. et Sasaki M. (1975). Location of nucleolar organizers in animal and plant chromosomes by means of an improved N-banding technique. *Chromosoma*, 49, 357–370.

Gabriel-Robez O., Jaafar H., Ratomponirina C., Boscher J., Bonneau J., Popescu C.P. et Rumpler Y. (1988). Heterosynapsis in a heterozygous fertile boar carrier of a 3;7 translocation. *Chromosoma*, 97, 26–32.

Gagné R. et Laberge C. (1972). Specific cytological recognition of the heterochromatic segment of number 9 chromosome in man. *Exp. Cell Res.*, 73, 239–242.

Gall J.G. (1991). Organelle assembly and function in the amphibian germinal vesicle. *Adv. Developmental Biochemistry*, 1, 1–29.

Gall J.G., Murphy C., Callan H.G. et Wu Z. (1991). Lampbrush chromosomes. *Methods Cell Biol.*, 36, 149–166.

Gallien L., Labrousse M., Picheral B. et Lacroix J.C. (1965). Modifications expérimentales du caryotype chez un Amphibien Urodèle (*Pleurodeles waltlii* Michah.) par irradiation de l'oeuf et la greffe nucléaire. *Rev. Suisse Zool.*, 72, 59–85.

Gardiner K. (1995). Human genome organization. *Curr. Opin. Genet. Dev.*, 5, 315–322.

Gardiner K. (1996). Base composition and gene distribution: critical patterns in mammalian genome organization. *Trends Genet.*, 12, 519–523.

Gasser S.M. et Laemmli U.K. (1987). A glimpse at chromosomal order. *Trends Genet.*, 3, 16–22.

Gatti M., Pimpinelli S. et Santini G. (1976). Characterization of Drosophila heterochromatin. I. Staining and decondensation with Hoechst 33258 and quinacrine. *Chromosoma*, 57, 351–375.

Gatti M., Bonaccorsi S. et Pimpinelli S. (1994). Looking at Drosophila mitotic chromosomes. *Meth. Cell Biol.*, Academic Press, 44, pp 371–391.

Geffrotin C., Popescu C.P., Cribiu E.P., Boscher J., Renard C., Chardon P. et Vaiman M. (1984). Assignment of MHC in swine to chromosome 7 by in situ hybridization and serological typing. *Ann. Génét.*, 27, 213–219.

Geitler L. (1938). Chromosomenbau. Protoplasma. Monographie, 14, Berlin Borntraeger.

Gerhard D.S., Kawasaki E.S., Bancroft F.C. et Szabo P. (1981). Localization of a unique gene by direct hybridization in situ. *Proc. Natl. Acad. Sci.*, 78, 3755–3759.

Goodpasture C. et Bloom S.E. (1975). Visualization of nucleolar organizer regions in mammalian chromosomes using silver staining. *Chromosoma*, 53, 37–50.

Goureau A., Yerle M., Schmitz A., Riquet J., Milan D., Pinton P., Frelat G. et Gellin J. (1996). Human and porcine correspondence of chromosome segments using bidirectional chromosome painting. *Genomics*, 36, 252–262.

Gray J.W., Carrano A.V., Steinmetz L.L., Van Dilla M.A., Moore D.H., Mayall B.H. et Mendelsohn M.L. (1975). Chromosome measurement and sorting by flow systems. *Proc. Natl. Acad. Sci.*, 72, 1231–1234.

Gray J.W., Peters D., Merrill J.T., Martin R. et Van Dilla M.A. (1979). Slit-scan flow cytometry of mammalian chromosomes. *J. Histochem. Cytochem.*, 27, 441–444.

Gray J.W. et Langlois R.G. (1986). Chromosome classification and purification using flow cytometry and sorting. *Ann. Rev. Biophys. Chem.*, 15, 195–235.

Greenbaum I.F., Gunn S.J., Smith S.A., McAllister B.F., Hale D.W., Baker R.J., Engstrom M.D., Hamilton M.J., Modi W.S., Robbins L.W., Rogers D.S., Ward O.G., Dawson W.D., Elder F.F.B., Lee M.R., Pathak S. et Stangl Jr. F.B. (1994). Cytogenetic Nomenclature of deer mice, *Peromyscus* (Rodentia): revision and review of the standardized karyotype. Report of the Committee for the Standardization of Chromosomes of *Peromyscus*. *Cytogenet. Cell Genet.*, 66, 181–195.

Grunwald D., Geffrotin C., Chardon P., Frelat G. et Vaiman M. (1986). Swine chromosomes: Flow sorting and spot blot hybridization. *Cytometry*, 7, 582–588.

Guest W.C. et Hsu T.C. (1973). A new technique for preparing Drosophila neuroblast chromosomes. *Drosophila Information Service*, 50, 193.

Gustavsson I. et Settergren I. (1984). Reciprocal chromosome translocation with transfer of centromeric heterochromatin in the domestic pig karyotype. *Hereditas*, 100, 1–5.

Haar F.M., Durm M., Aldinger K., Celeda D., Hausmann M., Ludwig H. et Cremer C. (1994). A rapid FISH technique for quantitative microscopy. *Biotechniques*, 11, 346–353.

Habeebu S.S.M., Spathas D.H. et Ferguson-Smith M.A. (1990). Non-radioactive in situ hybridization of DNA probes to chromosomes and nuclei. *Mol. Biol. Med.*, 7, 423–435.

Hagman B. (1993). Fluoreszenzoptische Untersuchungen zur Optimierung der Chromosomenpräparation. Diploma-Thesis, Institute of Applied Physics, University of Heidelberg.

Harper M.E. et Saunders G.F. (1981). Localization of single copy DNA sequences on G-banded human chromosomes by in situ hybridization. *Chromosoma*, 83, 431–439.

Harris P., Cooke A., Boyd E., Young B.D. et Ferguson-Smith M.A. (1987). The potential of family flow karyotyping for the detection of chromosome abnormalities. *Hum. Genet.*, 76, 129–133.

Hartley S.E. et Horne M.T. (1985). Cytogenetic techniques in fish genetics. *J. Fish Biol.*, 26, 575–582.

Hausmann M. (1984). Laserfluoreszenzaktivierte Analyse und Sortierung von Metaphasechromosomen, Anwendung der Slit-Scan Flussphotometrie. PhD Thesis, Faculty of Natural Sciences, University of Heidelberg, Allemagne.

Hausmann M., Dudin G., Aten J.A., Heilig R., Diaz E. et Cremer C. (1991). Slit-scan flow cytometry of isolated chromosomes following fluorescence hybridization: an approach of on-line screening for specific chromosomes and chromosome translocations. *Z. Naturforsch.*, 46c, 433–441.

Hausmann M., Dölle J., Arnold A., Stepanow B., Wickert B., Boscher J., Popescu C.P. et Cremer C. (1992). Development of a two-parameter slit-scan flow cytometer for screening of normal and aberrant chromosomes: application to a karyotype of *Sus scrofa domestica* (pig). *Opt. Eng.*, 31, 1463–1469.

Hausmann M., Popescu C.P., Boscher J., Kerboeuf D., Dölle J. et Cremer C. (1993). Identification and cytogenetic analysis of an abnormal pig chromosome for flow cytometry and sorting. *Z. Naturforsch.*, 48c, 645–653.

Hausmann M., Dölle J., Schurwanz M. et Cremer C. (1995). Slit-scan flow fluorometry and sorting of chromosomes: a fast preanalysis system for microscopy. *Microsc. Analys.*, 7, 27–29.

Hausmann M., Dölle J. et Cremer C. (1996a). Slit-scan Durchflusszytometrie von Chromosomenaberrationen: Perspektiven in der Biologischen Dosimetrie. *Z. Med. Phys.*, 6, 59–67.

Hausmann M., Wickert B., Vogel M., Schurwanz M., Dölle J., Wolf D., Aldinger K. et Cremer C. (1996b). Optics and experimental resolution of the Heidelberg slit-scan flow fluoremeter. *Proc. S.P.I.E.*, 2629, 146–156.

Hayes H., Petit E. et Dutrillaux B. (1991). Comparison of RBG-banded karyotypes of cattle, sheep and goats. *Cytogenet. Cell Genet.*, 57, 51–55.

Hayes H. (1995). Chromosome painting with human chromosome-specific DNA libraries reveals the extent and distribution of conserved segments in bovine chromosomes. *Cytogenet. Cell Genet.*, 71, 168–174.

Hediger R. (1988). Die in situ Hybridisierung zur Genkartierung beim Rind und Schaf. Abhandlung. Diss. ETH. Nr 8725.

Heitz E. et Bauer H. (1933). Beweise für die Chromosomennatur der Kernschleifen in der Knäuelkernen von Bibio hortulans. *L. Z. Zellforsch. Mikrosk. Anat.*, 17, 67–82.

Hess O. (1966). Structural modifications of the Y chromosome in *Drosophila hydei* and their relation to gene activity. *Chromosomes Today*, 1, 197–173.

Holmquist G. (1975). Hoechst 33258 fluorescent staining of Drosophila chromosomes. *Chromosoma*, 49, 333–356.

Holmquist G. (1977). C-banding, depurination and beta-elimination. *J. Cell Biol.*, 75, 140a.

Holmquist G.P. (1989). Evolution of chromosome bands: molecular ecology of noncoding DNA. *J. Mol. Evol.*, 28, 469–486.

Holmquist G.P. (1992). Review article: chromosome bands, their chromatin flavors, and their functional features. *Am. J. Hum. Genet.*, 51, 17–37.

Holmquist G., Gray M., Porter T. et Jordan J. (1982). Characterization of Giemsa dark- and light-band DNA. *Cell*, 31, 121–129.

Iannuzzi L. (coordinator) (1994). Standard karyotype of the river buffalo (*Bubalus bubalis L.*, 2n = 50): Report of the committee for the standardization of banded karyotypes of the river buffalo. *Cytogenet. Cell Genet.*, 67, 102–113.

ISCN (1978). An International System for Human Cytogenetic Nomenclature: Birth defects. Original Article Series, Vol. 14, N° 8, The National Foundation, New York 1978; also in *Cytogenet. Cell Genet.*, 21, 309–404.

ISCN (1981). An International System for Human Cytogenetic Nomenclature-High Resolution Banding: Birth Defects. Original Article Series, Vol. 17, N° 5, March of Dimes Birth Defects Foundation, New York, 1981; also in *Cytogenet. Cell Genet.*, 31, 1–23.

ISCN (1995). An international system for human cytogenetic nomenclature, F. Mitelman (ed), Karger, Basel.

ISCNDA (1989). The Second International Conference on Standardization of Domestic Animal Karyotypes, Di Berardino D., Hayes H., Fries R., Long S.E. (eds). *Cytogenet. Cell Genet.*, 53, 65–78, (1990).

Jaafar H., Gabriel-Robez O., Ratomponirina C., Boscher J., Bonneau M., Popescu C.P. et Rumpler Y. (1989). Analysis of synaptonemal complexes in two fertile heterozygous boars, both carriers of a reciprocal translocation involving an acrocentric chromosome. *Cytogenet. Cell Genet.*, 50, 220–225.

Jantch M., Hamilton B., Mayr B. et Schweizer D. (1990). Mitotic chromosome behaviour reflects levels of sequence divergence in *Sus scrofa domestica* satellite DNA. *Chromosoma*, 88, 330–338.

Jaylet A. (1967). Accidents chromosomiques obtenus à l'état hétérozygote dans la descendance de mâles irradiés, chez le triton *Pleurodeles waltlii* Michah. *Cytogenetics*, 6, 390–401.

Jaylet A. (1972). Tétraploïdie expérimentale chez le Triton *Pleurodeles waltlii* Michah. *Chromosoma*, 38, 173–184.

Jauch A., Wienberg J., Stanyon R., Arnold N., Tofanelli S., Ishida I. et Cremer T. (1992). Reconstruction of genomic rearrangements in great apes and gibbons by chromosome painting. *Proc. Natl. Acad. Sci.*, 89, 8611–8615.

John H.A., Birnstiel M.L. et Jones K.W. (1969). RNA-DNA hybrids at the cytological level. *Nature*, 223, 234–238.

Kallioniemi A., Kallioniemi O.-P., Sudar D., Rutovitz D., Gray J.W., Waldman F. et Pinkel D. (1992). Comparative genomic hybridization for molecular cytogenetic analysis of solid tumors. *Science*, 258, 818–821.

Kessler C. (1990). The digoxygenin system: principle and applications of the novel nonradioactive DNA labeling and detection system. *Bio Technology Int.*, 183–194.

King R.C. (1965). Genetics. Oxford University Press, New York.

King R.C. (1975). *Drosophila melanogaster*: an introduction. In: Handbook of Genetics, King, R.C. (ed), Plenum Press, New York, 3, pp 625–652.

Knoll M. (1992). Fluoreszenzoptishe Chromosomenanalyse einer permanenten Zellinie. Diploma-Thesis, Institute of Applied Physics, University of Heidelberg.

Kokalj-Vokac N., Almeida A., Viegas-Péquignot E., Jeanpierre M., Malfoy B. et Dutrillaux B. (1993). Specific induction of uncoiling and recombination by azacytidine in classical satellite-containing constitutive heterochromatin. *Cytogenet. Cell Genet.*, 63, 11–15.

Korenberg J.E. et Rykowski J.R. (1988). Human genome organization: Alu, Lines, and the molecular structure of metaphase chromosome bands. *Cell*, 53, 391–400.

Labrousse M. (1966). Analyse des effets des rayonnements appliqués à l'oeuf sur la structure caryologique et sur le développement embryonnaire de l'amphibien urodèle *Pleurodeles watlii* Michal. *Bull. Soc. Zool. France*, 91, 491–588.

Labrousse M. (1970). Topographie des chromosomes mitotiques larvaires chez deux espèces d'Urodèles. *C.R. Acad. Sci. (Paris)*, 271, 2134–2136.

Labrousse M. (1971). Sur la localisation et la transmission d'une mutation chromosomique viable chez l'Amphibien urodèle: *Pleurodeles waltlii* Michah. *Chromosoma*, 33, 409–420.

Lacroix J.C. (1968a). Etude descriptive des chromosomes en écouvillon dans le genre *Pleurodeles* (Amphibien, Urodèle). *Ann. Embr. Morph.*, 1, 179–202.

Lacroix J.C. (1968b). Variations expérimentales ou spontanées de la morphologie et de l'organisation des chromosomes en écouvillon dans le genre *Pleurodeles* (Amphibien, Urodèle). *Ann. Embr. Morph.*, 1, 205–248.

Lacroix J.C. (1970). Mise en évidence sur les chromosomes en écouvillon de *Pleurodeles poireti* Gervais, Amphibien Urodèle, d'une structure liée au sexe, identifiant le bivalent sexuel et marquant le chromosome W. *C. R. Acad. Sci. (III)*, 271, 102–104.

Lacroix J.C., Azzouz R., Boucher D., Abbadie C., Pyne C.K. et Charlemagne J. (1985). Monoclonal antibodies to lampbrush chromosome antigens of *Pleurodeles waltlii*. *Chromosoma*, 92, 69–80.

Lacroix J.C., Azzouz R., Simon F., Bellini M., Charlemagne J. et Dournon C. (1990). Lampbrush W and Z heterochromosome characterization with a monoclonal antibody and heat-induced chromosomal markers in the newt *Pleurodeles waltl*: W chromosome plays a role in female sex determination. *Chromosoma*, 99, 307–314.

Lacroix J.C. et Loones M.T. (1971). Fragmentation par les rayons X de l'organisateur d'une différenciation de chromosomes en écouvillon (Lampbrush), chez *Pleurodeles waltlii*. *Chromosoma*, 36, 112–118.

Lacroix J.C. et Loones M.T. (1974). Sélection par les chromosomes en écouvillon, de lignées vectrices de mutations chez l'Amphibien Urodèle *Pleurodeles poireti*. *Chromosoma*, 48, 297–326.

Ladjali K., Tixier-Boichard M. et Cribiu E.P. (1995). High resolution chromosome preparations for G- and R-banding in *Gallus domesticus*. *J. Hered.*, 86, 136–139.

Lalande M., Kunkel L.M., Flint A. et Latt S.A. (1984). Development and use of metaphase chromosomes flow sorting methodology to obtain recombinant phage libraries enriched for parts of human X chromosome. *Cytometry*, 5, 101–107.

Landegent J.E., Jansen in de Wal N., Baan R.A., Hoeijmakers J.H.J. et van der Ploeg M. (1984). 2-acetylaminofluorine modified probes for the indirect hybridocytochemical detection of specific nucleic acid sequences. *Exp. Cell Res.*, 153, 61–72.

Landegent J.E., Jansen in de Wal N., Dirks R.W., Baas F. et Van der Ploeg M. (1987). Use of whole cosmid cloned genomic sequences for chromosomal localization by non-radioactive in situ hybridization. *Hum. Genet.*, 77, 366–370.

Langer P.R., Waldrop A.A. et Ward D.A. (1981). Enzymatic detection of biotin labeled polynucleotides: novel nucleic acid affinity probes. *Proc. Natl. Acad. Sci.*, 78, 6633–6637.

Langlois R.G., Yu L.C., Gray J.W. et Carrano A.V. (1982). Quantitative karyotyping of human chromosomes by dual beam flow cytometry. *Proc. Natl. Acad. Sci.*, 79, 7876–7880.

Latt S.A. (1973). Microfluorometric detection of deoxyribonucleic acid replication in human metaphase chromosomes. *Proc. Natl. Acad. Sci.*, 70, 3395–3399.

Lau Y.F. et Arrighi F.E. (1977). Comparative studies of N-banding and silver staining of NORs in human chromosomes. In: Aspects of the chromosome organization and function, Drets M.E., Brum-Zorilla N. et Folle G.A. (eds), Monograph of Joint Seminar and Workshop, Montevideo, Uruguay, pp 49–55.

Lawrence J.B., Villnave C.A. et Singer R.H. (1988). Sensitive, high-resolution chromatin and chromosome mapping in situ: presence and orientation of two closely integrated copies of EBV in a lymphoma line. *Cell*, 52, 51–61.

Lebo R.V., Bruce B.D., Dazin P.F. et Payan D.G. (1987). Design and application of a versa-tile triple-laser cell and chromosome sorter. *Cytometry*, 8, 71–82.

Lee J-Y., Koi M., Stanbridge E.J., Oshimura M., Kumamoto A.T. et Feinberg A.P. (1994). Simple purification of human chromosomes to homogeneity using muntjac hybrid cells. *Nature Genet.*, 7, 29–33.

Lefevre G. Jr. (1976). A photographic representation and interpretation of the polytene chromosomes of *Drosophila melanogaster* salivary glands. In: The Genetics and Biology of Drosophila, Ashburner M. and Novitski E. (eds), Academic Press, 1a, pp 32–66.

Lemeunier F., Dutrillaux B. et Ashburner M. (1978). Relationships within the melanogaster subgroup species of the genus Drosophila (Sophophora). III. The mitotic chromosomes and quinacrine fluorescent patterns of the polytene chromosomes. *Chromosoma*, 69, 349–361.

Lemeunier F. et Aulard S. (1992). Inversion polymorphism in *Drosophila melanogaster*. In: Drosophila inversion polymorphism, Krimbas C.B. and Powell J.R. (eds), CRC Press.

Lemieux N., Dutrillaux B. et Viégas-Péquignot E. (1992). A simple method for simultane-ous R- or G-banding and fluorescence in situ hybridization of small single copy genes. *Cytogenet. Cell Genet.*, 59, 311–312.

Lenstra J.A., van Boxtel J.A.F., Zwaagstra K.A. et Schwerin M. (1993). Short interspersed nuclear element (SINE) sequences of the *Bovidae*. *Anim. Genet.*, 24, 33–39.

Lichter P., Cremer T., Borden J., Manuelidis L. et Ward D.C. (1988a). Delineation of indi-vidual human chromosomes in metaphase and interphase cells by in situ suppression hybridization using recombinant DNA libraries. *Hum. Genet.*, 80, 224–234.

Lichter P., Cremer T., Tang C-J.C., Watkins P.C., Manuelidis L. et Ward D.C. (1988b). Rapid detection of human chromosome 21 aberrations by in situ hybridization. *Proc. Natl. Acad. Sci.*, 85, 9664–9668.

Lichter P., Tang C.C., Call K., Hermanson G., Evans G.A., Housman D. et Ward D.C. (1990). High resolution mapping of human chromosome 11 by in situ hybridization with cosmid clones. *Science*, 247, 64–69.

Lichter P., Boyle A.L., Cremer T. et Ward D.C. (1991). Analysis of genes and chromosomes by nonisotopic in situ hybridization. *Genet. Anal. Techn. Appl.*, 8, 24–35.

Lin M.S., Latt S.A. et Davidson R.L. (1974). Microfluorometric detection of asymmetry in centromeric regions of mouse chromosomes. *Exp. Cell Res.*, 86, 392–395.

Lin M.S., Comings D.E. et Alfi D.S. (1977). Optical studies of the interaction of 4'-6-diamidino-2-phenylindole with DNA and metaphase chromosomes. *Chromosoma*, 60, 15–25.

Loones M.T. (1979). In vivo effects of g-irradiation on the functionnal architecture of the lampbrush chromosomes in Pleurodeles (Amphibia, Urodela). *Chromosoma*, 73, 357–368.

Lucas J. et Gray J.W. (1987). Centromeric index versus DNA content flow karyotypes of human chromosomes measured by means of slit-scan flow cytometry. *Cytometry*, 8, 273–279.

Lucas J., Mullikin J. et Gray J. (1991). Dicentric chromosome frequency analysis using slit-scan flow cytometry. *Cytometry*, 12, 316–322.

Luciani J.M., Morazzani M.R. et Stahl A. (1975). Identification of pachytene bivalents in human male meïosis using G-banding technique. *Chromosoma*, 52, 275–282.

Lyon M.F. (1962). Sex chromatin and gene action in the mammalian X chromosome. *Am. J. Hum. Genet.*, 14, 135–148.

Mahdy E.A., Mäkinen A., Chowdhary B.P., Andersson L. et Gustavsson I. (1989). Chromo-somal localization of the ovine major histocompatibility complex (OLA) by in situ hybridization. *Hereditas*, 111, 87–90.

Mäkinen A. (coordinator) (1985a). The standard karyotype of the silver fox (*Vulpes fulvus* Desm.). Committee for the standard karyotype of *Vulpes fulvus* Desm. *Hereditas*, 103, 171–176.

Mäkinen A. (coordinator) (1985b). The standard karyotype of the blue fox (*Alopex lagopus* L.). Committee for the standard karyotype of *Alopex lagopus L. Hereditas*, 103, 33–38.

Mäkinen A., Chowdhary B.P., Mahdy E., Andersson L. et Gustavsson I. (1989). Localization of the equine major histocompatibility complex (ELA) to chromosome 20 by in situ hybridization. *Hereditas*, 110, 93–96.

Malcolm S., Cowell J.K. et Young B.D. (1986). Specialist techniques in research and diagnostic clinical cytogenetics. In: Human cytogenetics. A practical approach, Rooney D.E. et Czepulkowski B.H. (eds), Oxford, England, IRL Press, pp 197–226.

du Manoir S., Speicher M.R., Joos S., Schröck E., Popp S., Döhner H., Kovacs G., Robert-Nicoud M., Lichter P. et Cremer T. (1993). Detection of complete and partial chromosome gains and losses by comparative genomic in situ hybridization. *Hum. Genet.*, 90, 590–610.

Manuelidis L. (1985). Individual interphase chromosome domains revealed by in situ hybridization. *Hum. Genet.*, 71, 288–293.

Manuelidis L. et Ward D.C. (1984). Chromosomal and nuclear distribution of the HindIII 1.9 kb human DNA repeat segment. *Chromosoma*, 91, 28–38.

Massaad L., Venuat A.M., Luccioni C., Beaumatin C., Lemieux N. et Dutrillaux B. (1991). High catabolism of BrdU may explain unusual sister chromatid differenciation and replication banding patterns in cancer cells. *Cancer Genet. Cytogenet.*, 53, 23–34.

Medrano L., Bernardi G., Couturier J., Dutrillaux B. et Bernardi G. (1988). Chromosome banding and genome compartmentalization in fishes. *Chromosoma*, 96, 178–183.

Melamed M.R., Lindmo T. et Mendelsohn M.M. (eds) (1990). Flow cytometry and sorting. Wiley-Liss, New York.

Metz C.W. (1914). Chromosome studies in the Diptera. I. A preliminary survey of five different types of chromosome groups in the genus Drosophila. *J. exp. Zool.*, 17, 45–59.

Metz C.W. (1916a). Chromosome studies in the Diptera. II. The paired association of chromosomes in the Diptera and its significance. *J. exp. Zool.*, 21, 213–279.

Metz C.W. (1916b). Chromosome studies in the Diptera. III. Additional chromosome groups in the Drosophilidae. *Amer. Nat.*, 50, 587–599.

Mezzelani A., Castiglioni B., Eggen A. et Ferretti L. (1996). T-banding pattern of bovine chromosomes and karyotype reconstitution with physically mapped cosmids. *Cytogenet. Cell Genet.*, 73, 229–234.

Miller O.L. et Beatty B.R. (1969). Visualization of nucleolar genes. *Science*, 164, 955–957.

Miller O.J., Miller D.A., Dev V.G., Tantravahi R. et Croce C.M. (1976). Expression of human and suppression of mouse nucleolus organizer activity in mouse-human somatic cell hybrids. *Proc. Natl. Acad. Sci.*, 73, 4531–4535.

Moore K.L. et Barr M.L. (1955). Smears from the oral mucosa in the detection of chromosomal sex. *Lancet*, ii, 57.

Moorhead P.S., Nowell P.C., Mellman W.J. et Battips D.M. (1960). Chromosome preparations of leucocytes cultured from human peripheral blood. *Exp. Cell Res.*, 20, 613–616.

Moses M.J. (1977). The synaptonemal complex and meiosis. In: Molecular human cytogenetics, Sparkes R.S., Comings D. et Fox C.F. (eds), New York, Academic Press, vol. II, pp 101–125.

Nederlof P.M., van der Flier S., Wiegant J., Raap A.K., Tanke H.J., Ploem J.S. et van der Ploeg M. (1990). Multiple fluorescence in situ hybridization. *Cytometry*, 11, 126–131.

Okada T.A. et Comings D.E. (1974). Mechanisms of chromosome banding. III. Similarity between G-bands of mitotic chromosomes and of chromomeres of meïotic chromosomes. *Chromosoma*, 48, 65–71.

Otto F.J., Oldiges H., Göhde W., Barlogie B. et Schumann J. (1980). Flow cytogenetics of uncloned and cloned Chinese hamster cells. *Cytogenet. Cell Genet.*, 27, 52–56.

Otto F. et Tsou K.C. (1985). A comparative study of DAPI, DIPI, and Hoechst 33258 and 33342 as chromosomal stains. *Stain Techn.*, 60, 7–11.

van den Oven C. et Aten J.A. (1990). Instrument for real-time pulse-shape analysis of slit-scan flow cytometry signals. *Cytometry*, 11, 630–635.

Painter T.S. (1933). A new method for the study of chromosome rearrangements and the plotting of chromosome maps. *Science*, 78, 585–586.

Paris Conference (1971), Supplement (1975). Standardization in human cytogenetics: Birth defects. Original Article Series, Vol. 11, N° 9, The National Foundation, New York 1975; also in *Cytogenetics*, 15, 201–238, (1975).

Pardue M.L. et Gall J.G. (1969). Molecular hybridization of radioactive DNA to the DNA of cytological preparations. *Proc. Natl. Acad. Sci.*, 64, 600–604.

Parra I. et Windle B. (1993). High resolution visual mapping of stretched DNA by fluorescent hybridization. *Nature Genet.*, 5, 17–21.

Penrad-Mobayed M., Bonnanfant-Jaïs M.L., N'Da E. et Angelier N. (1986). Evidence for a particular mode of transcription in globular loops of lampbrush chromosomes of the newt *Pleurodeles waltlii*. *Chromosoma*, 94, 319–328.

Perry P. et Wolff S. (1974). New Giemsa method for differential staining of sister chromatids. *Nature*, 261, 156–158.

Perry P. et Evans H.L. (1975). Cytological detection of mutagen exposure by sister chromatid exchange. *Nature*, 258, 121–125.

Pimpinelli S., Santini G. et Gatti M. (1976). Characterisation of Drosophila heterochromatin. II. C- and N-banding. *Chromosoma*, 57, 377–386.

Pimpinelli S., Berloco M., Fanti L., Dimitri P., Bonaccorsi S., Marchetti E., Caizzi R., Caggese C. et Gatti M. (1995). Transposable elements are stable structural components of *Drosophila melanogaster* heterochromatin. *Proc. Natl. Acad. Sci.*, 92, 3804–3808.

Pinkel D., Landegent J., Collins C., Fuscoe J., Segraves R., Lucas J. et Gray J. (1988). Fluorescence in situ hybridization with human chromosome-specific libraries: detection of trisomy 21 and translocations of chromosome 4. *Proc. Natl. Acad. Sci.*, 85, 9138–9142.

Popescu C.P. (1971). Les chromosomes méiotiques du boeuf. *Ann. Génét. Sél. Anim.*, 3, 125–143.

Popescu C.P., Bonneau M., Tixier M., Bahri I. et Boscher J. (1984). Reciprocal translocations in pig. Their detection and consequences on animal performance and economic losses. *J. Hered.*, 75, 448–452.

Popescu C.P. (coordinator), Long S.E., Riggs P., Womack J., Schmutz S., Fries R. et Gallagher D.S. (1996). Standardization of cattle karyotype nomenclature: Report of the committee for the standardization of the cattle caryotype. *Cytogenet. Cell Genet.*, 74, 259–261.

Pyne C.K., Loones M.T. et Lacroix J.C. (1989). Correlation between isolated lampbrush chromosomes and nuclear structures of *Pleurodeles waltl* oocytes: An electron microscopic study. *Chromosoma*, 98, 181–193.

Pyne C.K., Simon F., Loones M.T., Géraud G., Bachmann M. et Lacroix J.C. (1994). Localization of antigens PwA33 and La on lampbrush chromosomes and on nucleoplasmic structures in the oocyte of the urodele *Pleurodeles waltl*: Light and electron microscopic immunocytochemical studies. *Chromosoma*, 103, 475–485.

Pyne C.K., Loones M.T., Simon F. et Zhou Z.J. (1995). Immunocytochemical study of lampbrush chromosomes of the urodele *Pleurodeles waltl*: axial granules are recognized by the mitosis-specific monoclonal antibody MPM-2. *Biol. Cell*, 83, 191–200.

Rabbitts P., Impey H., Heppel-Parton A., Langford C., Tease C., Lowe N., Bailey D., Fergusson-Smith M. et Carter N. (1995). Chromosome specific paints from a high resolution flow karyotype of the mouse. *Nature Genet.*, 9, 369–375.

Ragghianti M., Bucci S., Mancino G., Lacroix J.C., Boucher D. et Charlemagne J. (1988). A novel approach to cytotaxonomic and cytogenetic studies in the genus *Triturus* using monoclonal antibodies to lampbrush chromosomes antigens. *Chromosoma*, 97, 134–144.

Reading Conference (1976). Proceedings of the first international conference for the standardization of banded karyotypes of domestic animals. Reading, England, 1976, Ford C.E., Pollok D.L., Gustavsson I. (eds). *Hereditas*, 92, 145–162, (1980).

Rens W., van Oven C.H., Stap J. et Aten J.A. (1993). Effectiveness of pulse-shape criteria for the selection of dicentric chromosomes by slit-scan flow cytometry and sorting. *Analyt. Cell Pathol.*, 5, 147–159.

Rens W., van Oven C.H., Stap J., Jakobs M.E. et Aten J.A. (1994). Slit-scanning technique using standard cell sorter instruments for analysing and sorting nonacrocentric human chromosomes including small ones. *Cytometry*, 16, 80–87.

Renshaw H.W., Eckblad P., Everson D.O., Tassinari P.D. et Amos D. (1977). Ontogeny of immunocompetence in cattle: evaluation of phytomitogen induced in vitro bovine foetal lymphocyte blastogenesis, using a whole blood culture technique. *Am. J. Vet. Res.*, 38, 1141–1150.

Rettenberger G., Lett C., Zechner U., Kunz J., Vogel W. et Hameister H. (1995a). Visualization of the conservation of synteny between humans and pigs by heterologous chromosomal painting. *Genomics*, 26, 372–378.

Rettenberger G., Klett C., Zechner U., Bruch J., Just W., Vogel W. et Hameister H. (1995b). ZOO-FISH analysis: cat and human karyotypes resemble the putative ancestral mammalian karyotype. *Chromos. Res.*, 3, 479–486.

Richard F., Vogt N., Muleris M., Malfoy B. et Dutrillaux B. (1994). Increased FISH efficiency using APC probes generated by direct incorporation of labeled nucleotides by PCR. *Cytogenet. Cell Genet.*, 65, 169–171.

Richer C.L., Power M.M., Klunder L.R., McFeely R.A. et Kent M.G. (1990). Standard karyotype of the domestic horse (*Equus cabalus*). Committee for standardized karyotype of *Equus cabalus*. The second International Conference for Standardization of Domestic Animal Karyotypes, INRA, Jouy-en-Josas, France, 22–26 May 1989. *Hereditas*, 112, 289–293.

Ried T., Baldini A., Rand T.C. et Ward D.C. (1992). Simultaneous visualization of seven different DNA probes by in situ hybridization using combinatorial fluorescence and digital imaging microscopy. *Proc. Natl. Acad. Sci.*, 89, 1388–1392.

Rigby P.W.J., Diechmann M., Rhodes C. et Berg P. (1977). Labeling deoxyribonucleic acid to high specific activity in vitro by nick translation with DNA polymerase I. *J. Mol. Biol.*, 113, 237–251.

Rodriguez-Martin M.L., Herberts C., Moreau N. et Angelier N. (1989). Effects of in vivo heat treatment on lampbrush chromosome structure in amphibians oocytes. *Exp. Cell Res.*, 185, 546–550.

Roth M.B. et Gall J.G. (1987). Monoclonal antibodies that recognize transcription unit proteins on newt lampbrush chromosomes. *J. Cell Biol.*, 105, 1047–1054.

Rudak E., Jacobs P.A. et Yanagimachi R. (1978). Direct analysis of the chromosome constitution of human spermatozoa. *Nature*, 274, 911–913.

Saccone S., De Sario A., Della Valle G. et Bernardi G. (1992). The highest gene concentrations in the human genome are in telomeric bands of metaphase chromosomes. *Proc. Natl. Acad. Sci.*, 89, 4913–4917.

Saccone S., De Sario A., Wiegant J., Raap A.K., Della Valle G. et Bernardi G. (1993). Correlation between isochores and chromosomal bands in the human genome. *Proc. Natl. Acad. Sci.*, 90, 11929–11933.

Saitoh Y. et Laemmli U.K. (1994). Metaphase chromosome structure: Bands arise from a differential folding path of the highly AT-rich scaffold. *Cell*, 76, 609–622.

van de Sande J.H., Lin C.C. et Jorgenson K.F. (1977). Reverse banding on chromosomes produced by a guanosine-cytosine specific DNA binding antibiotic: olivomycin. *Science*, 195, 400–402.

Schardin M., Cremer T., Hager H.D. et Lang M. (1985). Specific staining of human chromosomes in Chinese hamster X man hybrid cell lines demonstrates interphase chromosome territories. *Hum. Genet.*, 71, 281–287.

Scheres J.M. (1976). CT banding of human chromosomes. The role of cations in the alkaline pretreatment. *Hum. Genet.*, 33, 167–174.

Scherthan H., Cremer T., Arnason U., Weier H.-U., Lima de Faria A. et Frönicke L. (1994). Comparative chromosome painting discloses homologous segments in distantly related mammals. *Nature Genet.*, 6, 342–347.

Schmitz A., Oustry A., Chaput B., Bahri-Darwich I., Yerle M., Milan D., Frelat G. et Cribiu E.P. (1995). The bovine bivariate flow karyotype and peak identification by chromosome painting with PCR generated probes. *Mammal. Genome*, 6, 415–420.

Schnedl W. (1971). Analysis of the human karyotype using a reassociation technique. *Chromosoma*, 34, 448–454.

Schnedl W., Breitenbach M., Mikelsaar A.V. et Stranzinger G. (1977). Mithramycin and DAPI: a pair of fluorochromes specific for GC and AT rich DNA respectively. *Hum. Genet.*, 36, 299–305.

Schreck R.R., Erlanger B.F. et Miller O.J. (1974). The use of antinucleoside antibodies to probe the organization of chromosomes denatured by ultraviolet irradiation. *Exp. Cell Res.*, 88, 31–39.

Schultz L.D., Kay B.K. et Gall J.G. (1981). In vitro RNA synthesis in oocyte nuclei of the newt *Notophthalmus viridescens*. *Chromosoma*, 82, 171–187.

Schweizer D. (1976). Reverse fluorescent chromosome banding with chromomycin and DAPI. *Chromosoma*, 58, 307–324.

Schweizer D., Ambros P. et Anderle M. (1978). Modification of DAPI banding on human chromosomes by prestaining with a DNA-binding oligopeptide antibiotic, distamycin A. *Exp. Cell Res.*, 111, 327–332.

Scott S.E.M. et Sommerville J. (1974). Location of nuclear proteins on the chromosomes of newt oocytes. *Nature*, 250, 680–682.

Seabright M. (1971). A rapid banding technique for human chromosomes. *Lancet*, 2, 971–972.

Searle J.B., Fedyk S., Fredga K., Hausser J. et Volobouev V.T. (1991). Nomenclature for the chromosomes of the common shrew (*Sorex araneus*). *Mem. Soc. Vaud. Sc. Nat.*, 19.1, 13–22.

Sèle B., Pellestor F., Estrade C., Ostorero C., Warembourg E., Gelas M., Jalbert H. et Jalbert P. (1985). Mise en évidence des chromosomes de spermatozoïdes humains dans un système hétérospécifique: difficultés techniques. *Path. Biol.*, 33, 875–880.

Sharon N. et Lis H. (1989). Lectins. Chapman and Hall, London and New York.

Shoffner R.M., Krishan A., Haiden G.J., Bammi R.K. et Otis J.S. (1967). Avian chromosome methodology. *Poultry Sci.*, 46, 333–344.

Silber J., Bazin C., Lemeunier F., Aulard S. et Volovitch M. (1989). Distribution and conservation of the Foldback transposable element in Drosophila. *J. Mol. Evol.*, 28, 220–224.

Sillar R. et Young B. (1981). A new method for the preparation of metaphase chromosomes for flow analysis. *J. Histochem. Cytochem.*, 29, 74–78.

Simi L.B., Sasi R., Lingrel J.B. et Lin C.C. (1989). Mapping of the goat beta-globin gene cluster to a region of chromosome 7 by in situ hybridization. *J. Hered.*, 80, 246–249.

Singer M. (1982a). Highly repeated sequences in mammalian genomes. *Int. Rev. Cytol.*, 76, 67–112.

Singer M.F. (1982b). SINEs and LINEs: highly repeated short and long interspersed sequences in mammalian genomes. *Cell*, 28, 433–434.

Solinas-Toldo S., Lengauer C. et Fries R. (1995). Comparative genome map of human and cattle. *Genomics*, 27, 489–496.

Sorsa V. (1988a). Polytene chromosomes in genetic research. Wiseman A. (ed), Ellis Horwood Limited, Chichester.

Sorsa V. (1988b). Chromosome maps of Drosophila. Vol. II. CRC Press, Boca Raton, Florida.

Southern E.M. (1975). Detection of specific sequences among DNA fragments separated by gel electrophoresis. *J. Mol. Biol.*, 98, 503–517.

Stanyon R., Arnold N., Koehler U., Bigoni F. et Wienberg J. (1995). Chromosomal painting shows that "marked chromosomes" in lesser apes and Old World monkeys are not homologous and evolved by convergence. *Cytogenet. Cell`Genet.*, 68, 74–78.

Stefos K. et Arrighi F.E. (1971). Heterochromatin nature of the W chromosome. *Exp. Cell Res.*, 68, 228–231.

Stevens N.M. (1912). The chromosomes of *Drosophila ampelophila*. *Carnegie Inst. Washington Publ.*, 36, 380–381.

Stöhr M., Hutter K., Frank M. et Futterman G. (1980). A flow cytometric study of chromosomes from rat kangooroo and Chinese hamster cells. *Histochem.*, 67, 179–190.

Stubblefield E., Cram S. et Deaven L. (1975). Flow microfluorometric analysis of isolated Chinese hamster chromosomes. *Exp. Cell Res.*, 94, 464–468.

Stubblefield E. et Oro J. (1982). Isolation of specific chicken macrochromosomes by zonal centrifugation and flow sorting. *Cytometry*, 2, 273–281.

Sumner A.T. (1972). A simple technique for demonstrating centromeric heterochromatin. *Exp. Cell Res.*, 75, 304–306.

Sumner A.T. (1989). Chromosome banding – Animals. *Genome*, 31, 467–468.

Sumner A.T., Evans H.J. et Buckland R.A. (1971). A new technique for distinguishing between human chromosomes. *Nature*, 232, 31–32.

Switonski M., Reimann N., Bosma A., Long S.E., Barnitzke S., Pienkowska A., Moreno-Milan M.M. et Fischer P. (1996). Committee for the Standardized Karyotype of the Dog (*Canis familiaris*): Report on the progress of the standardization of the G-banded canine (*Canis familiaris*) karyotype. *Chromos. Res.*, 4, 306–309.

Szabo P. (1989). In situ hybridization. In: Human Chromosomes. Manual of basic techniques, Verma R.S. et Babu A. (eds), New York, USA, Pergamon Press, pp 152–165.

Takahashi E., Hori T., O'Connell P., Leppert M. et White R. (1990). R-banding and non-isotopic in situ hybridization: precise localization of the human type 11 collagen gene (COL2A). *Hum. Genet.*, 86, 14–16.

Tarkowski A.K. (1966). An air-drying method for chromosome preparations from mouse eggs. *Cytogenetics*, 5, 394–400.

Tchen P., Fuchs R.P.P., Sage E. et Leng M. (1984). Chemically modified nucleic acids as immunodetectable probes in hybridization experiments. *Proc. Natl. Acad. Sci.*, 81, 3466–3470.

Trask B., van den Engh G., Mayall B. et Gray J. (1989). Chromosome heteromorphism qualified by high resolution bivariate flow karyotyping. *Am. J. Hum. Genet.*, 45, 739–752.

Trask B. (1991). Fluorescence in situ hybridization. *Trends Genet.*, 7, 149–154.

Vassart M., Séguéla A. et Hayes H. (1995). Chromosomal evolution in gazelles. *J. Hered.*, 86, 216–227.

Viegas-Péquignot E. et Dutrillaux B. (1976). Segmentation of human chromosomes induced by 5-ACR (5-azacytidine). *Hum. Genet.*, 34, 247–254.

Viegas-Péquignot E. et Dutrillaux B. (1978). Une méthode simple pour obtenir des prophases et des prométaphases. *Ann. Génét.*, 21, 122–125.

Viegas-Péquignot E., Dutrillaux B., Magdelenat H. et Coppey-Moisan M. (1989). Mapping of single-copy DNA sequences on human chromosomes by in situ hybridization with biotinylated probes: enhancement of detection sensitivity by intensified-fluorescence digital-imaging microscopy. *Proc. Natl. Acad. Sci.*, 86, 582–586.

Vogel W., Autenrieth M. et Speit G. (1986). Detection of bromodeoxyuridine incorporation in mammalian chromosomes by a bromodeoxyuridine-antibody. I. Demonstration of replication patterns. *Hum. Genet.*, 72, 129–132.

Wahl G.M., Stern M. et Stark G.R. (1979). Efficient transfer of large DNA fragments from agarose gels to diazobenzylozymethyl paper and rapid hybridization by using dextran sulfate. *Proc. Natl. Acad. Sci.*, 76, 3683–3687.

Wallace R.A., Jared D.W., Dumont J.N., Sega M.W. (1973). Protein incorporation by isolated amphibian oocytes. III. Optimum incubation conditions. *J. Exol. Zool.*, 184, 321–333.

Ward D.C. (1990). New developments in non-isotopic labeling and purification of nucleic acid probes. *Clin. Biochem.*, 23, 307–310.

Weier H.-U.G., Wang M., Mullikin J.C., Zhu Y., Cheng J.-F., Greulich K.M., Bensimon A. et Gray J.W. (1995). Quantitative DNA fiber mapping. *Hum. Mol. Genet.*, 4, 1903–1910.

Weisblum B. et de Haseth P.L. (1972). Quinacrine, a chromosome stain specific for deoxyadenylate-deoxythymidylate rich regions in DNA. *Proc. Natl. Acad. Sci.*, 69, 629–632.

Wheeless L.L. et Patten S.F. (1973). Slit-scan cytofluorometry: Basis for an automated cytopathology prescreening system. *Acta Cytol.*, 17, 391–394.

Wheeless L.L. (1990). Slit-scanning. In: Flow cytometry and sorting, Melamed M.R., Lindmo T. et Mendelsohn M.L. (eds), Wiley-Liss, New York.

Wiegant J., Ried T., Nederlof P.M., van der Ploeg M., Tanke H.J. et Raap A.K. (1991). In situ hybridization with fluoresceinated DNA. *Nucl. Ac. Res.*, 19, 3237–3241.

Wienberg J., Stanyon R., Jauch A. et Cremer T. (1992). Homologies in human and Macaca fuscata chromosomes revealed by in situ suppression hybridization with human chromosome specific DNA libraries. *Chromosoma*, 101, 265–270.

Willard H.F. et Waye J.S. (1987). Hierarchical order in chromosome-specific human alpha satellite DNA. *Trends Genet.*, 3, 192–198.

Wu Z., Murphy C., Callan H.G. et Gall J.G. (1991). Small Nuclear Ribonucleoproteins and Heterogeneous Nuclear Ribonucleoproteins in the Amphibian Germinal Vesicle: Loops, Spheres, and Snurposomes. *J. Cell Biol.*, 113, 465–483.

Xeros N. (1962). Deoxyriboside control and synchronization of mitosis. *Nature*, 194, 682–683.

Yanagimachi R., Yanagimachi H. et Rogers B.J. (1976). The use of zona-free animal ova as a test-system for the assessment of the fertilizing capacity of human spermatozoa. *Biol. Reprod.*, 15, 471–476.

Yerle M., Schmitz A., Milan D., Chaput B., Monteagudo L., Vaiman M., Frelat. G. et Gellin J. (1993). Accurate characterization of porcine bivariate flow karyotype by PCR and fluorescence in situ hybridization. *Genomics*, 16, 97–103.

Yunis J.J. (1976). High resolution of human chromosomes. *Science*, 191, 1268.

Index

11-dUTP fluorescein   70
11-dUTP-biotin   70, 73, 74, 180, 185
11-dUTP digoxigenin   70, 74, 180
16-dUTP-biotin   74, 146
5-azacytidine   57
5-azadeoxycytidine   57
5-methylcytosine rich-bands   43, 44, 46
5-methylcytosine   43, 44

A
A ribonuclease   75
Abnormal chromosomes   85
Acetic carnoy   125
Acetic violet   4, 6
Acridine orange   22, 41, 42, 151
Alpha satellite or alphoid repetitive DNA
   27, 58, 62
Alu sequence   28, 31
Amethopterin   8, 10, 55
Aminopterin   8
Amphibia (chromosomes)   27, 103, 124
Amphibian chromosomes   27, 103, 124
Aneuploidy   205
Anoure   103
Anti-5-methylcytosine antibodies   43, 63
Anti-biotin antibody   70, 79
Anti-BrdU antibody   48, 49
Anti-digoxygenin antibody   184, 185 (••)
AT alignement   28, 29, 30
Autoradiography   69, 77
Avidine   70
Axial granules (lampbrush chromosomes)
   129
Axial structures (lampbrush
   chromosomes)   128

B
B lymphocytes   1
B1 sequence   31
Background   77, 169
Bands   7, 25, 27, 28, 29, 30

Barr body   151, 152
Bibionidés   141
Biopsies   3
Bird (avian) chromosomes   12, 27, 54
Bird (avian) female heterogamety   12
Bird (cells culture)   5
Bird (chromosomes)   5, 13, 27, 31, 54, 103
Bird chromosome sexing   12, 13
Black light   49
Blastula   151
Blood (culture)   1, 5, 14
Bone marrow cells   12
Borate (buffer solution)   202
Bovine (chromosomes)   52, 53
BrdU incorporation   12
BrdU incorporation asymmetry   56
BrdU photolyse   10, 15, 49
Bromodeoxyuridine (BrdU)   10, 15
BSS+ (buffer solution)   201
Buffer solutions (protocol for preparation)
   201

C
Carabidae   17
C-bands   60
CBG-bands   14, 33, 58
Cebus   61
Cell cycle   7, 47
Cell harvest   10
Cells count   4, 5
Cells thawing   7
Centromere   26
Centromeric index (*CI*)   172, 173
Centromeric region   57, 58
Cerambycidae   17
Cercopithecide   62
Cerebroid ganglions (Drosophila)   138
Chimpanzee   62
Chironomidae   141, 142
Chromatin   26
Chromatine halo   152

Chromomere   129
Chromomycin A3   189
Chromosome banding   25
Chromosome magnetic separation   182
Chromosome painting   80
Chromosome rearrangement   80
Chromosome sorting   161
Chromosome spreads   18
Chromosomes denaturation   76
Chromosomes in suspension   178
Chromosomes nomenclature   XV, XVI
Chrysomelidae   17
Colcemid   9
Colchicine   9
Coleoptera   16
Collembole   16
Compaction (DNA)   25
Comparative *in situ hybridisation*   80
Competitor (DNA)   80
Concanavalin A (Con A)   3, 62
Constitutive heterochromatin   60, 58, 26
Corona radiata   96
Cryoprotector agent   7
CT-bands   61
Culture medium   197
Cumulus-oocyte complex   95
Cytofluorometry   157

D
DAB (diaminobenzidine)   197
DA-DAPI   62
DAPI   53, 62
Daunomycin   29
Defollicularisation   114
Denhardt (solution)   196
Depellucidation (oocytes)   100
Derm cells   3
Dextran sulphate   196
Diapause   16
Dicentric chromosomes   176
Differential renaturation   30
Dimethylsulfoxide (DMSO)   7
Diptera   17
Distamycin A   187
DNA denaturation   72
DNA satellite   59
DNA synthesis inhibition   9
Double synchronisation   8
Drosophila   17
Drosophila ampelophila   137

Drosophila melanogaster   143
Dynabeads   185
Dynamic bands   31

E
Earle 10X (buffer solution)   202
Early replication   27
EDTA   18
Electron microscopy   92, 120
Eliomys quercinus   60
Ethanol-acetic acid fixation   19
Ethidium bromide   163, 166
Euchromatin   26

F
Fading   51
Female meiosis   94
Fibroblasts culture   3
Ficoll   2
Fish chromosomes   13
Fixation   19
"Flavor" model   28
Flow cytometry   157
Flow karyotype   158
Flow sorting   161
Fluorescein (FITC)   52, 71
Fluorodeoxyuridine (FdU)   8
Foetal cells   3
Foetal ovary   94
Foetus   4
Follicle   94
Formamide   191
Formamide deionisation   191
FPG (staining)   49

G
G11-bands   61
GAG-bands   36
Gazelle (chromosomes)   36
G-bands   34
GBG-bands   58
GBP-bands   51
Genetic variant (lampbrush chromosomes matrices)   107
Genomic DNA (preparation from blood)   190
Germicide lamp   44
Germinal vesicle   96, 102
Giant chromosome   143
Giemsa   20, 58
Giemsa R   20
Goat (chromosomes)   60

Golden hamster (*Mesocricetus auratus*)
100
Gorilla 62
GTG-bands 34

H
HCG (Human Chorionic Gonadotropin)
100
Heterochromatin 61, 137
Heterologous chromosome painting 80
Heterospecific fertilisation 97
Heterosynapsis (pig) 94
Hipotonic (solution) 192
Histone 58
Hoechst 33258 188
Human (chromosomes) 21, 42
Hyaluronidase 191
Hybridisation of chromosomes in
suspension 178, 185
Hybridised chromosomes identification
71
Hypotonic treatment 18

I
Immunolabelling (lampbrush
chromosomes) 130
Immunodetection 111
Immunomorphologic labelling
(lampbrush chromosomes) 111
Immunostaining 48
In situ hybridisation 69, 153
Insect chromosomes 16, 17
Interbands 141
Interphasic nucleus 153
Interspecific in vitro fertilisation 100
Isochore 28

J
Juxtacentromeric 61

K
Karyotype XV
Kinetochore 26

L
L1 sequence 28
Lampbrush chromosomes 103
Lampbrush chromosomes anti-protein
antibody 111
Lampbrush map 112
Laser 159, 160
Late replication 26

Lectine 1
Lemur coronatus 61
LINEs 28
Lipopolysaccharide (LPS) 14
Liquid nitrogen 7
Loops (lampbrush chromosomes) 104,
105
Loops matrices (lampbrush
chromosomes) 105, 129, 130
Lorisidae 60
Lung cells 4
Lymphoblastic transformation 1
Lymphocytes (culture) 1, 2
Lymphocytes culture 1
Lymphocytes separation 2

M
M structures (lampbrush chromosomes)
119, 129
Mac Ilvaine (buffer solution) 202
Macrochromosome 12
Magnetic beads 185
Male meiosis 85
Male sterility 85
Mallasez chamber (cells count) 4
Meiosis 85
Meiotic chromosomes 85
Meiotic division 85
Mesocricetus auratus 100
Methanol-acid acetic fixation 19
Methotrexate 10
Methylation 41
Microchromosome 27
MIF-1 sequence 31
Mithramycin 31
Mitogen 1
Mitotic cells 1, 2
Mitotic cells accumulation 18
Mitotic index 2, 14
Mitotic spindle 10
Monomorphic specie 12
Morphologic variant (lampbrush
chromosomes matrices) 107
Musca domestica (chromosomes) 120

N
Nick translation labelling 73
Non histone protein 25, 58
Non radioactive *in situ* hybridisation 70
Non radioactive labelling (probes) 73
Non radioactive probe 73
NOR (staining) 65

Nucleolus   27
Nucleolus organising region   65
Nycticebus coucang   60

O
Olivomycin   31
Oocyte   114
Orang-outang   62
Orcein   21
Ovaire foetal   95
Ovaries biopsies (amphibia)   115

P
Paraformaldehyde   117
PBS⁻ (buffer solution)   192
Polymerase chain reaction   74
Perodicticus potto   60
Peroxydase   197
Phosphate (buffer solution)   203
Photographic emulsion   77
Photolyse du BrdU   49
Phytohaemagglutinin (PHA)   1
Pig (chromosomes)   41, 43, 178, 179
Plathyrhini   61
Pleurodeles poireti (lampbrush
    chromosomes)   104
Pleurodeles waltlii (lampbrush
    chromosomes)   104
Pleurodele   103
Pleurodele (mitotic chromosomes)   103
PMSG (Pregnant Mare Serum
    Gonadotropin)   100
Pokeweed mitogene (PWM)   1, 2
Polar body   96
Polytene chromosome   137, 141
Polyténisation   141, 142
Pongidae   62
Porc (chromosomes)   93
PPD (p-phenylenediamine)   51, 79
Prefixation   19
Primary culture   2
Primate   26, 61
Probe (labelling)   72, 73
Prometaphase cells   7
Prometaphase chromosome   7
Propidium iodide   24
Prosimien   60
Proteinase K   154
Puffs   142

Q
Q-bands   29
QFQ-bands   34

Quinacrine   188
Quinacrine mustard   188

R
Radioactive *in situ* hybridisation   75
Radioactive labelling   75
Radioactive probe   73
RBA-bands   23
R-bands   27, 30
RBG-bands   50
RBP-bands   24, 32
RHG-bands   37, 39
Ringer (solution)   203
Robertsonian fusion   60

S
S phase (cell cycle)   27, 54
Salivar gland   138
SAR (Scaffold Associated Region)   28
SAR region   28
Scarabaeidae   17
Secondary constriction   57
Sequential chromosome banding   65
Sex bivalent (pig)   93
Sex chromatin   151
Sheep (chromosomes)   32, 34, 39
Sheep anti-rabbit IgG fluoroscein
    conjugated antibodies   79
Silver nitrate   93
Simple synchronisation   8
SINEs   28
Sister chromatids exchanges   56
"Slit-scan" cytometry   172
"Slit-scan" profil of chromosomes   173,
    174, 176
Snurposom   106
Sodium diatriazoate
Sodium heparin   2
Sorensen (buffer solution)   201
Sorted chromosomes (identification)
    176
Spermatocyte   88
Spermatogenesis   88
Spermatogenetic cycle   8
Spermatogony   85
Spermatozoid   97
Spermatozoid chromosomes   87
Spermatozoids capacitation   99
Spermiogenesis   98
SSC20X (buffer solution)   201
Standardisation   XV, XVI
Stock solutions   199
Structural bands   30

Synaptonemal complex 86
Synchronisation 8

T
T bands 39, 42
T lymphocytes 1
TBS (buffer solution) 203
TE10X (buffer solution) 202
Telomer 26
Telomeric 26
Telomeric region 26
Testis biopsy 88
Thymidine 203
Trisomy 21 94
Trypan blue 6
Trypsin 18
Trypsination

U
Urodeles 103

W
W chromosome 14, 136

X
X chromosome 54, 87
X chromosome inactivation 87

Y
Y chromosome 56, 62, 64
YAC 53

Z
ZOO-FISH 80

Printing (Computer to Film): Saladruck, Berlin
Binding: Stürtz AG, Würzburg